# DEEP IMMERSION
## The Experience of Water

As Father Teilard de Chardin is to the noosphere and Gaston
Bachelard is to reverie, Robert France is to water—an oracle of
aqua. With the mastery of a Carlos Castaneda and the passion
of a Joan Halifax demanding that the world change, this
unabashed nature lover with a Thoreauvian intelligence and a
sweeping frame of reference delivers a pagan-positive message:
embrace whatever we most value in nature. Through us the
world becomes able to see itself at last . . . a wonderful book!

> ✌ Lewis MacAdams, founder of Friends of the Los
> Angeles River and author of *The River: Books 1 & 2*

I've often wondered why I'm attracted to water—why I like to
sit by it, contemplate it, plunge into it. *Deep Immersion* tells me
more about my aquatic obsession than any other book I've ever
read. But Robert France's book is not just about water. It's also
about our unhealthy, indeed our perishing environment: I rec-
ommend it to anyone who cares about life on this beleaguered
planet.

> ✌ Lawrence Millman, explorer and author of numerous
> books including *Last Places* and *Lost in the Arctic*

There are few more important disconnections to heal in today's
world than humanity's rift with life-giving water. *Deep Immersion*
is rich with insight into water's fundamental place in culture,
history, the human spirit, and our individual souls. It offers a
deep well for thirsty minds.

> ✌ *Sandra Postel*, director of the Global Water Policy
> Project and author of *Last Oasis* and *Pillar of Sand: Can
> the Irrigation Miracle Last?*

The directionality of moving water is powerful, yet creates a peace. With water we can visit the world, learn to move with it, and be part of its moments. Through vivid depictions of the tactile experience of water, Robert France reminds us of the spiritual senses that proximity to water can awaken. Just as a knot in water loosens and untangles with time, our hearts and minds can likewise be loosened and untangled by the presence of water. Frozen, still, flowing, or as vapor in clouds, water has the ability to heighten the focus of the human spirit. *Deep Immersion* offers us—as no other book has—instruction in how to engage our planet's water. Whether dipped into here and there or plunged into completely, this book celebrates a hymn to humankind's grand aquatic dance. For as Thales—Thoreau's favorite philosopher—wrote over two millennia ago: "Everything is water; water is all."

      John Middendorf, professional river guide, international big wall climber, equipment designer and author of *Big Walls*

These essays eloquently capture the arguments against hard-ened, technical solutions to 'manage' water—they convincingly steer the reader toward connecting and interacting with water in its determined fluid form and unconstrained boundaries. This is a wonderful and inspirational book—It should be prerequisite for all students entering any learning program associated with studying, modifying, or for that matter living in the natural world. My, I've just said that everyone should read it!

      Carolyn Adams, director of the Watershed Science institute of the Natural Resources Conservation Service

What can one say about *Deep Immersion*, whether in print or in life? As Robert France draws upon a flood of creative writings in this book, it seems apt to compare his project with that of another distinguished scientist, Yi-Fu Tuan, whose writings cascaded from arid zone geomorphology to *The Hydrologic Cycle* and *Wisdom of God, Topophilia*, and beyond. But while Tuan turned from a focus on *philia* to more turbulent emotional questions, France offers an exuberant *aquaphilia* for those who are or might become lovers of water.

> ✎ James L. Wescoat, Jr., head of the Department of Landscape Architecture at the University of Illinois and coauthor of *Water for Life: Water Management and Environmental Policy* and *Mughal Gardens: Sources, Places, Representations, and Prospects*

Running deep through us, water is an element essential to life. Acknowledging the biological and spiritual life energies that embrace us like the two banks of a river is easy; capturing or providing that essential experience to enable or reinvigorate the connection and dependency we have with water is much harder. In his comprehensive and fully integrated synthesis, Harvard's Robert France has provided us with a deep and thoughtful mind-body experience of the elemental strands that bind us to water. No one who reads *Deep Immersion* can remain untouched by the fundamental need to reconnect with our aqueous heritage. This book will not only expand your experience of the world to more fully include the interconnection between earth's elements and our souls, but also why the need to cross over between disciplines is necessary to gain a fuller appreciation of life itself.

> ✎ Mark Chandler, international conservation director at Earthwatch Institute

"In *Deep Immersion*, Robert France immerses us in water, its meaning and history, its literature and legacy, is spirituality and ecopsychology. Redolent of Thoreau, the book tells us how water is used, abused, and wasted; its importance to all life and to life processes. *Deep Immersion* also tells us what we must do to protect water as our most precious resource. France evokes all things ever said or thought or written about water, and offers us an exhaustive bibliography for yet further immersions on our own."

                ∽        Verne Huser, environmental mediator and author/
                        editor of *River Reflections: A Collection of River Writings*
                        and *Down by the River: The Impact of Federal Water Projects
                        & Policies on Biological Diversity*

**ROBERT LAWRENCE FRANCE, Ph. D.** is one of the world's leading environmental scholars and is a professor at the Harvard Design School where he teaches courses on water-sensitive ecological planning and design as well as environmental theory. Dr. France has authored over a hundred technical papers and authored or edited numerous technical books in addition to serving as supervising editor for his own series at Lewis Publishers/CRC Press entitled *Integrated Studies in Water Management and Land Development*. He is also the editor of *Reflecting Heaven: Thoreau on Water* (Houghton Mifflin, 2001), *Profitably Soaked: Thoreau's Engagement with Water* (Green Frigate Books, 2003) and *Water-Logged-In: A Dynamic Library of Aquatic Quotations from Thoreau's Descendants*. A complete listing of his publications and current research interests can be found at the webpage gsd.harvard.edu/faculty/france.

**LAURA SEWALL, Ph. D.** is trained in the psychology and neurophysiology of vision and is a professor at Prescott College in Arizona where she teaches courses on ecopsychology. One of the pioneers in this developing field, Dr. Sewall has published several papers and is the author of the book *Sight and Sensibility: The Ecopsychology of Perception* (Jeremy P. Tarcher/Putnam, 1999).

**HERBERT DREISEITL** is an internationally renowned landscape architect whose many design projects incorporate and focus on water. Mr. Dreiseitl is co-author of the book *Waterscapes: Planning, Building and Designing with Water* (Birkhauser Press, 2001).

# DEEP IMMERSION
## The Experience of Water

Robert Lawrence France

Forewords by
Laura Sewall
and by
Herbert Dreiseitl

## GREEN FRIGATE BOOKS
Sheffield, Vermont

Library of Congress Control Number 2002090980
ISBN 0-9717468-1-8

First Edition
Printed in United States of America

Green Frigate Books,
P.O. Box 461, Sheffield, VT
U.S.A. 05866-0461
www.greenfrigatebooks.com

France, Robert Lawrence
Deep Immersion: The Experience of Water
Forewords by Laura Sewall; Herbert Dreiseitl
p. cm.
LCCN: 2002090980
ISBN: 0-9717468-1-8
1. Water. 2. History. 3. Ecopsychology. 4. Ecocriticism.
5. Landscape Architecture. 6. France, Robert.

Cover photograph is by Linda Troeller from her book Healing Waters (troeller@bway.net)

# Dedication

To those who seek to reconnect us with imagination and passion, hope and love, to the world of water from which we all sprang. To the swimming monkeys of Nagano, primates who have had the courage to return, if but only sporadically . . . so far. To the late Victor Hugo Daza, slain by the Bolivian army at a peaceful protest against the foreign privatization of water. To the Australian soldiers who were shot and killed while swimming and playing in the beautiful ocean surf as a reprieve from the horrors of the battle for Gallipoli. And especially to L.S., for the briefest of immersions.

*To experience water . . . is to be cleansed by it, to see its beauties, and understand more deeply than ever before its manifold values.*

> —James Wescoat, Jr. "Beneath which rivers flow: Water, geographic imagination and sustainable landscape design," 2002

*Take thought,*
*when you are speaking of water,*
*that you first account your experiences,*
*and only after your reflections.*
—Leonardo da Vinci

*The union of the sensory and the sensual*
*upholds a moral value. The contemplation*
*and experience of water leads us, by many*
*routes, to an ideal.*
—Gaston Bachelard

# Contents

# Foreword One

*The activity of thinking is essentially an expression of flowing movement . . . With this ability to enter thoughtfully into everything and to picture all things in the form of ideas, the process of thinking partakes in the laws of the formative processes of the universe. These are the same laws at work in the fluid element, which renounces a form of its own and is prepared to enter into all things, to unite all things, to absorb all things.*
—Theodor Schwenk, *Sensitive Chaos: The Creation of Flowing Forms in Water and Air*, 1978

*The life of the mind is not the rotation of a machine through a cycle of fixed phases, but the flow of a torrent through its mountain-bed, scattering itself in spray as it plunges over a precipice and pausing in the deep transparency of a rockpool.*
—R . G. Collingwood, *The Idea of Nature*, 1945

# A Fluidity of Mind

*Laura Sewall*

I live at the mouth of an estuary, where fresh water spills into salt water, and where the ocean fingers its way into land. It's a place of privilege. I know this because of the gratitude I awaken with each day. In my mind, privilege is the opportunity to be shaped by tides, and by earthy sounds, and by an early morning flood of light. I've also come to know that one's heart cannot help but follow the goodness of such penetrating influence. The headlines in today's paper catch my eye. The first reads: "Do Society's Yardsticks Really Measure Up?" A caption under a photo of the Nasdaq Market site in Times Square adds: "U.S. society is quantifying everything today, from the stock market to school scores."

The second headline reads: "It's a drought—no, a flood!" The byline follows, "What in the world is going on with weather?" The first of several photos shows a father and son high and dry in a rowboat, caught in a dry lake bed in Beijing—caught in the worst drought of either's lifetime. A second photo shows a woman battling through flood waters in downtown Buenos Aires. Headline two illustrates water problems: the increasingly devastating droughts, over consumption, and the damage to human settlement caused by floods linked to deforestation and climate change.

But what of the first headline? What does the socialization of numbers—the cultural contract with quantification—have to do with either ecopsychology or water?

The lenses of ecopsychology are finely focused on the interface between where we are—our environments, nature, our homes, the places we live and love—and the state of our psyches. It is the relationship, a radical form of psychological and spiritual inter-dependence with the natural order of all things, that is the center of ecopsychological attention.

The collective psyche, meaning the one conditioned by common, cul-turally-determined experience, is of particular interest. The collective psyche is constituted by parts of our consciousness and behavior that are mutually reinforced by our everyday interactions. For example, most of us share the myriad experiences derived from the quantification that saturates our culture: dollar amounts signify value; days partitioned into hours and minutes determine our embodied experience, or how we move through each day; and now, performance in school—for twelve formative years of reinforced training—is to be increasingly ranked. Apparently, we are training ourselves to diminish the realm of possibility to nominal scales with consequently reduced ranges of values; and thus, to objectify reality.

There is a certain sensibility that arises when trained on the objectifi-cation and minimization of reality. It is detached, not quite involved. It tends to be competitive, and most interested in facts, in knowing for sure. It thus lacks imagination and an appreciation for the unknown, and for the numinous. I also suspect that such a sensibility casts our attention into the distance, far from the immediacy and intimacy of the tangible world. We then want a "good view" or "greener grass." Considering our numbers, such a sensibility wrecks havoc on the planet.

We design our places. We make our consumer choices. But with what sensibility do we do so? And what, pray tell, most informs and conditions such a sensibility? Like a dose of the very best medicine, two times a day, my senses are entrained by the easing in and out of water. Conditioned by the tides, my senses are tuned to a pulse, to a ceaseless and subtle cadence that stretches and shapes my world—that touches an ancestral knowledge of rhythmic reality, a natural truth. Each day, I witness the soft slippage of water, in and out like a long, slow breath. A similarly slow ratcheting in my mind associates the tides with the fact of being feminine, so pulled and tugged by the Moon. For a moment, like a glint of mirror, I see myself in the tidal pulse—also rhythmic, also changing, also vast. This, I'm grateful to say, is embodied knowledge, informing my thoughts and my every move.

I no longer speak of static realities. Like nouns, excessive quantification or absolute conclusions no longer interest me. My attention is drawn to the qualities—a curvature of shadow, the shimmer of light, the lilt of voice. My world is enriched by the qualities I hear and see, "out-there." Looking there, my pleasure deepens, and the knowing in my body shifts. No doubt, my actions follow—ever-more integrated, soft and ceaseless, and without need of distraction.

My hope for this volume is a contemporary revelation of Thoreau's embodied wisdom. Given the current state of water in the world, the "troubled waters," the drying up and polluting of rivers and lakes, and of our oceans, we need policy-makers, activists, and a citizenry deeply informed and enlisted by beauty, and by the fluidity and inspiration of water.

> *Cast your eyes, I beseech of you, upon the edge of water, the place where wild growth is undeterred, where a celebration of life arise from uncontained, untamed flow forms, where we see water as the fluid giver, the very source of all life, where lush, green beauty reigns!*

# Foreword Two

*Thus, a truly life-related type of thinking would find reflection in a corresponding landscape treatment. The straightening of rivers and other life-arteries would cease in favor of returning them to a properly meandering course. Thinking restored to a capacity to take in living formative impulse would be flowing thought and therefore prove hospitable to the flowing, formative forces inherent in systems of life-related laws. Within us, flowing wisdom; outside us, in nature, a water-element that is again allowed to be the carrier of wisdom-permeated laws, a water-element that works as a regulator in landscapes restored to a healthy state of being.*
—Theodor Schwenk, *Water, the Element of Life*, 1989

*Anyone who hikes through a hot woodland or on a rocky mountain trail experiences the same urge upon arriving at a stream, pool, or waterfall that a city worker experiences upon discovering a cool, splashing urban fountain after trudging the hot sidewalks on a summer day. The natural urge is to proceed beyond the visual and auditory experience: to look and listen is often not sufficient to satisfy our deepest needs. We want to feel the water in order to complete our experience.*
—Craig Campbell, *Water in Landscape Architecture*, 1978

# Waterworks

*Herbert Dreiseitl*

How does water integrate other elements? The beautiful phenomenon of water always surrounds us. Morning, evening and day colors are always influenced by the presence of water, the drops in the air refracting the light. Though frequently regarded from a perspective of physics, we should always look upon water as an art form, a poem. We need to understand the wisdom of what water means in life and how it supports life.

When we look at what water is, we always see reflections, or what water does. We can only see water in different places: as when it is frozen hard like stone or when it is as weightless vapors in the sky. We can therefore never see water itself.

Also we can never take water. It always disappears. Water is therefore very selfless. It is an element that gives itself, such that we can do almost anything with this water. When we look at water in nature it is certainly an element that integrates a lot of things. It is artist, it is landscape architect, it is also in a certain way, engineer. Always water has enough space and time in nature. No matter where we look, our landscapes—the shapes and forms—are very much influenced by water. Water is the ultimate landscape architect. We always see the borderlands between air and water and the shapes that water leaves behind such as the wave patterns in sand after the tides have gone out. There is an enormous potential of creativity in this water. No one snowflake is equal to another. Varied and dynamic, water is always changing. It lives in the beauty and artistry of little details such as a falling raindrop. And when we look at close-ups of water, we see that it is so magic and full of surprises. The magnified structure of water even resembles the early forms of life itself.

People can discover the importance of water in pristine nature as easily as by lifting a stone in a clear mountain stream and watching the small organisms flee in the current. Water is always trying to hide, and nowhere does it do this more successfully than in cities. What then do children in our cities feel about water? What experiences do they have? Mostly they don't know where it comes from or to where it goes. They have no responsibility to it. They know water only intellectually. They have no feeling for it.

Water in nature is characterized by soft sinusoid forms; in cities, by straight hard edges. Rivers now look like our roads: straight and quick. We have taken away time for water. We have taken water out of our society. Yet inside, we still depend on water and love it.

There is no need to get back to pristine nature even if we could. On weekends many people sit in plastic and metal boxes to go for two or three hours to have a sandwich or barbecue in "nature" and then return to their so called "no nature." Why do we have a civilization where our dreams are so far away? Why can't we work in a way that our living space has a quality that this is our new nature—our urban nature?

Problems always help us to wake up. The problems with water in cities—pollution and flooding—highlight the need to find a new way to manage water. Remarkable changes in attitude have occurred in this regard in the last twenty years.

In historical times, people were always strongly connected to water. They had a soul and spiritual and social connection with the element. People used to have no water in their homes and instead had to establish community around a central supply fountain. Water was celebrated in these early cultures with structures such as the cisterns integrated into the architecture of Moorish buildings in Spain. In all cases, small solutions showed a certain care and respect for the water environment.

To experience water in such an intimate way, even today, is to be quiet, to be fresh, to feel that something special is going on. People now need to rediscover old ways and to create new ways in which to experience water—to meditate around it, to listen to the noise it makes, the way it drowns out that which is not important.

# Prologue

*We attribute to water virtues that are antithetic to the ills of a sick person. Man projects his desire to be cured and dreams of a compassionate substance.*

                    —Gaston Bachelard, *Water and Dreams: An Essay on the Imagination of Matter*, 1999

# Aponiptein
# Dikaion Touton

*He would never know that such a simple act would become the single most profound demonstration of deep immersion in history, its resonance echoing and amplifying throughout the ages, burying what might have once been fact behind a veil of myth. He waited anxiously as his servant returned with the clay bowl of water, drawn no doubt from the palace cistern, itself fed by the very aqueduct that he had built for the ungrateful citizens. As his fingers entered the cool water, he could immediately feel the refreshing sensation against his skin. We wonder if he closed his eyes for a brief moment in concentration, ignoring the din of the clamoring crowd outside, and the startled attendants about him, and of course, the deafening silence of the prisoner at his feet. Was he able to succeed in his struggle to find peace? It was all he wanted. Could he will his fingers to draw up the serenity of the liquid? It must have lasted for only an instant, if that. For this immersion was not for personal hedonism, but was an act of ritual for the benefit of the crowd. Or was it?*

*As he moved his hands about in the water, rubbing one against the other, moving the cool liquid between his fingers, he must have thought about the reality behind the symbolism. He must have. In his head he knew it to be nothing more than a symbol, an act of contrition—aponiptein—"to wash off from." Their religion believed it: the redemptive power of such washing. Their prophets had written about it. "Our hands have not shed this blood," had said one. "Wash yourself, make yourself clean, put away evil from your souls . . . and though your sins be scarlet, I will make them white as wool; and if they be red like crimson, I will make them white as snow," had said another.*

*He continued, looking up only once to register that he had caught the fixed attention of the suddenly quiet crowd. They of course believed in the act. But the most intriguing question is whether he himself believed it. We will never know and can only surmise about what he might have hoped for in his heart. Certainly given all that had gone on up to that point—if we are to accept the written account—one might assume that he was deeply troubled by his role in the whole affair.*

*His own tradition likewise held that immersion in water was the gateway to purity. Had not a general recently cried out for the spirit of the Emperor in heaven to "come to these soldiers of yours and wash clean this stain," after putting down a mutiny? Like all magistrates, he had had water poured over his hands prior to performing sacrifices to the gods. We can only guess about how much he desperately wanted to believe that the same process of ablution would work here as well. How, after those unnerving looks into the prisoner's eyes, could he go on otherwise? In his heart he must have wished for it; this against, however, everything his reason and intellect told him.*

*How scared was he at that moment when he realized that there was no way out? To live with a sin of such magnitude must have been a frightening prospect indeed. For it is here where tradition suggests that yet another layer of personal tragedy came into play. For one assumes that surely he must have remembered the words of his famous countrymen: the orator who said that "a stain on the soul cannot vanish for a long time, not to be washed away by any rivers"; or the poet who wrote that "it's too easy to suppose . . . that those sad crimes can be carried away by water."*

*History only recounts the actions, leaving us to ponder the bigger mysteries of what their effects might have been upon the participants. Possibly fearing his own future due to the inescapable guilt, or possibly looking forward to getting on with his future free of any residual concerns, the water having done its job, he called for a towel from a servant and dried his hands "before the multitude, saying, 'I am innocent of the blood of this just person: see ye to it.'" This "just person," he had said, or "this innocent person"—*dikaion touton.

*The lingering question is whether the brief dipping of hands into the water had been enough in his mind to cleanse himself of the sin he recognized he was committing. Was he too a "dikaion touton"? Or did he return to the events with a troubled mind again and again over the succeeding years and wonder whether all would have been different in his life had he but had the opportunity to immerse his entire body deep into cleansing water? As if somehow it was the magnitude of the act of immersion that would have made all the difference. And perhaps—if we are to believe the myths—this might explain his end, purposefully walking into the lake, losing himself finally in those deep, comforting waters . . .*

# Introduction

*One can first of all simply wonder at the fact that it is only recently that humanity has begun to ponder the evolution and fate of water in the world when the very survival of our species depends upon it.*

—Lord Selborne, *The Ethics of Freshwater Use: A Survey*, 2000

*It would be a great victory over the machine if we were able to dispense with the increasingly complicated mechanization of our day and prove that we could get, at less expense, all that we needed from natural elements.*

—Jacques Cousteau, *The Silent World*, 1953

*There is, indeed, no single quality of the cat that man could not emulate to his advantage.*

—Carl Van Vechten, *in The Literary Cat*, 1990

# Ripple Effects

## Going Downhill

Things appear to have become so bad that it now seems impossible to consider the celebration of water as an art form in isolation of concerns about the protection of water as a threatened element. Even a coffee-table book such as *The Beauty and Mystery of Water*, that includes spectacular haunting images by photographer Hans Silvester, is bracketed by text from Bernard Fishesser and Marie-France Dupluis-Tate about the dismal current state of affairs. "Water is in a bad way," one sentence succinctly and apply states. This is due to the harsh reality, another sentence informs us, that "in its march toward the future, human society has disrupted the future of water."

Perhaps we should not be surprised. Since the very beginning, one of nature's truisms has always been that, left to itself, water runs downhill. Today unfortunately, not only is this a direct feature of the corporeal nature of water, it is also a symbolic feature of the status of that water. What if instead, the world operated as portrayed in Escher's wonderful *Waterfall* etching? Here, water falling from the top of a canal strikes a waterwheel before entering the bottom of the same canal, where by optical trickery it makes its way uphill through a series of Z-bend chicanes before impossibly reaching the top of the waterfall once more. Intrigued by this concept, the organizers of two recent confer-

ences used this image of Escher's as a metaphor for how human ingenuity can be drawn upon for improving the physicality of water sensitive
planning and design. But all involved in promoting the well-being of
water realize that merely focusing on technical solutions is, by itself, not
enough. How then do we similarly go about uplifting the symbolic status of water in our lives?

Given recognition of the growing scarcity and vulnerability of the
global water supply, UNESCO recently called for an urgent need "to
devise a common ethos" about water. Much debate is occurring in
international circles about how we should regard water. Is access and use
of potable water a need? Or is it instead a right? If water is to be looked
upon as a need, then it can be a commodity. And if it is a commodity,
then it can be owned and profit made from its sale and distribution. But
if instead, water is conceived of as an inalienable right for all to share, it
becomes not only impossible for anyone to profit from it, it also
becomes immoral. In April 2000, a seventeen-year-old Bolivian student,
Victor Hugo Daza, while at a large protest against water privatization by
foreign multinationals, was shot in the face by an army sniper, dying
instantly.

Even worse is the fact that for many, water is no longer regarded as
"water," but rather as a substance called "$H_2O$." The philosopher Ivan
Illich pessimistically reflected on the sad fact that what was once a fluid
that "drenches the inner and outer spaces of imagination," has now
become an industrial and technical utility, often corrosive to the skin. As
a culture, we have lost our collective memory of water as the archetypal
source of our dreams, replacing it with a water of "stuff," a mere
resource to be managed. Instead, argues Robert Wetzel, General
Secretary of the International Society of Limnology, we need to regard
freshwater as a common trust of humanity, requiring civilization "to
move united, assertively, and with dispatch to save the element, and thus
us all."

To foster a new morality of water, it is important, states UNESCO,
that we recognize that "water is not only a means to other goals, it is
essential as an end in itself." In other words, we need nothing short of
the development of a new water culture. Too long have we looked upon
water in pejorative terms as a harbinger of disasters due to scarcity, pollution, or conflict. Instead we need to hark to the closing words of the

Director-General of UNESCO in his message on the occasion of World Water Day 2000: "There is a fundamental truth which I would like to emphasize . . . the water supply does not run dry when it is drawn from the well of human wisdom."

The root of the problem lies in the sad reality that there has been an extinction of the experience of water in our lives that has left us both psychologically and spiritually desiccated. And like raisins, we have become the withered shells of our former turgid glory.

This book advances the simple thesis that it is only by truly experiencing water deeply and intimately that we can advance toward establishing a global consciousness about how to treat water respectfully and ethically. For, as Theodor Schwenk so aptly wrote "Now how do we go about re-enlivening water? There is no other way than to develop what professionals themselves describe as a 'new water consciousness.'" Such thoughts are by no means new. One of the earliest admirers of water's manifold lessons, Lao Tzu, wrote that "the sage's transformation of the world arises from solving the problem of water."

### Nature Designed—Simulated Experiences

We are too clever for our own good, and for that of the planet; too clever by far. Modern society, driven by arrogance, goes out of its way to regulate the environment and control nature. While real nature succumbs, we invent ever more elaborate artificial natures. Free from the annoying variances of real nature, we revel in our cleverness at being able to design the most appropriate environmental conditions to suit our every whim.

Blind to hubris, in thrall to technology, we play god to yearnings to experience water, whether they take the form of downhill skiing (during summer) on a giant façade of a mountain slope built within a warehouse near Tokyo, or surfing atop a giant curling wave (during winter) as it races toward a fake beach inside a shopping mall in Edmonton. Why go outside and experience nature itself in all its uncomfortable messiness when we can play on the tidy, unthreatening simulacra of our own cleverness? For that is what it really is: play. And like ignorant children, we confound and confuse the invented worlds of play with the troubling worlds of reality. Seduced by the former, we ignore the latter. Marveling as we enter, for example, The Biodome at the old Olympic Park site in

Montreal, we are greeted by a palette of mini-hydrological displays, becoming lulled into forgetting about the degraded St. Lawrence River that rushes near by, unheeded.

Eventually losing connection with the real water world outside, we become accustomed to, and think nothing strange or disturbing about paying money to consume fizzy bottled water from a distant continent while that in our neighborhood streams have become so unsafe that they have to be fenced off from us. Not that we really want to go there anyway, given the option of being able to experience water indirectly, through technology. "Contact" for many these days means nothing more than having constant access to a cell-phone. This distancing from the real world of water reached its apogee with the opening of a building at the EXPO 2002 site in Switzerland. We love this sort of stuff, so much so in fact, that we couldn't even wait until construction, the building design having been presented to admiring audiences at an exhibition in New York City and at an invited lecture at the Harvard Design School.

The "Blur Building" at the EXPO site is indeed impressive. But it is also seductively dangerous. It is designed to resemble a cloud floating over a lake, with the experience being one of entering a habitable medium "that is featureless, depthless, scaleless, massless, surfaceless, and contextless," rather than a traditional building. Indeed, the concept is to figuratively blur that which is natural with that which is unnatural. Rather than visiting the lake itself, it is the "building" that provides a simulation of such a visit.

An artificial cloud will be produced by 12,500 high-pressure water nozzles fixed to a 100 by 65 meter steel frame. A created weather system hooked into a computer will regulate the nozzles to produce more or less fog on demand, depending on bothersome climatic vagaries of temperature, humidity, and wind speed and direction. Therefore, rather than individuals adjusting to changing weather conditions, it will be the building that registers such shifts. Visitors will enter the cloud from a boardwalk and, once inside, will be met with "an overwhelming experience of sounds, sights and smells of the atomized lake water." There will be an opportunity for individuals to respond to the environment, but even here it is to be modulated through technology. Upon entering the building, visitors don obligatory "(b)raincoats" that are covered with sensors that can be digitally coded to personal preferences and that produce blush-like responses to the mist. Pervading all will be a sense of celebration at our cleverness in being able to reinvent nature. Forgotten is the real lake, fully capable of influencing ambient weather and producing its own fog and mist . . . naturally.

### Nature Desired—Reality Therapy

We desperately need to rediscover models and metaphors of instruction about how to re-experience the world of water about us. For as Susan Clifford and Angela King state "At the very moment when we need the closeness of water to feed our humanity and imagination, we seem to be denied literal contact, and have lost sight and sound of its magic." Although it might seem a strange analogy given their near uniform disdain for water, we could learn much about how to experience the world by observing the way in which cats go about it. For cats, even those who are house bound, life is a grand adventure. Fueled by imagination, they scamper about chasing this or that. It is no accident that their passion for exploration has a well-acknowledged risk associated with it, some felines succumbing due to their overt curiosities. For cats, too, life is a game to be played at with abandon. Frolicking, rolling, flying, they launch themselves into the environment, turning even the simplest act from the mundane to the magical. Cats are also the most sensual of creatures that we humans can easily observe. Not only do they seem to desire bodily contact with their world, rubbing their heads and tails against all they encounter, purring in bliss at the physicality of the act, but they actually seem to need such activity. And finally, cats seem to understand the rewards of contemplative downtime better than most. If the mass of men lead lives of quiet desperation, as Henry David Thoreau has intoned, those of cats lead lives of quiet contemplation.

Cats, then, have much to teach us about how to intimately interact with and deeply experience the world: **adventure, joy, contact, contemplation**—these provide the key. Thoreau knew this well. His two-million-word-plus *Journal* is filled with careful observations analyzing the behaviors of his various cats over the years. And it is bittersweet and telling to read his *Journal* entries for the last months of his life as he lay on his deathbed, his once expansive world of nature now shrunk to that of observing the patterns of water droplets traced upon the windowpane, his only experience of the world outside vicariously filtered through the reactive behavior of a kitten as she perched upon the windowsill. This, then, in the end is what it all came down to for the master phenomenologist: nature distilled to water and cats, complemented by ranging dreams of sailing and of native Indians in their canoes.

### Streams of Consciousness

This is a book about water; not so much about what it is, though there is some of that present, but rather, what it *does*. Specifically, what water does to those individuals who allow themselves to enter into a close relationship with it. *Deep Immersion: The Experience of Water* provides an overview of how to establish water—human connections by anchoring readers to a deeper understanding of the important roles played by water in all our lives. The idea, always swirling just beneath the surface throughout the book, is that the more we learn about and experience water, the more we are moved to give ourselves over to it—to deeply and profoundly immerse ourselves in its physical essence and spiritual nature, and in so doing, reciprocate by preserving its presence in our enriched lives, untainted.

The core of the book is formed by Part II, Chapters 4 through 7. Here for the first time is an exploration of the techniques used by contemporary nature writers concerning their direct engagement with water in lakes, rivers, wetlands, springs, ephemeral pools, and the ocean. Bracketing this *eco-criticism* section are a series of essays that provide the context in which to place the water writings of these worthy descendants of Thoreau. Part I reinforces the overwhelming importance of water in history, religion, literature, cinema, music, art, and architecture (Chapter 1), aquatic pollution and water use (Chapter 2), and deep ecology and hydrotherapy (Chapter 3)—what in other words, might be referred to as the *ecopsychology* of water. Part III provides examples of implementing these approaches through innovative environmental writing and ecological restoration (Chapter 8), experiential education and environmental art and performance (Chapter 9), and a lesson by way of Thoreau's deepest immersion (Chapter 10)—in other words, what could be conceived of as, in the broadest sense in terms of fostering nature connectivity, the *landscape architecture* of water. In both Parts I and III, the emphasis is clearly on introducing readers to a diverse array of scholarly fields, each more than worthy of an entire volume of its own. All essays are therefore accompanied by a large listing of references that will allow interested readers to more deeply immerse themselves in these topics.

### *Moving from Words to Action: Personally Applying the Message*

Reading a book, no matter how inspiring, is no substitute for direct experience. Thoreau certainly recognized this, and devoted as much time to actively exploring and deeply immersing himself in nature as he did for reading and writing about it. In an open letter to readers of *Orion Magazine*, several nature writers (some of whose work is introduced later in these pages) offered the following sage advise: "Words on a page do not accomplish anything by themselves; but words taken to heart, carried in mind, may lead to action." A major message of this book is that we need to do a much better job of living through our bodies, not just through our minds. For as Saint Teresa stated "The important thing is not to think much but to love much; and so do that which best stirs you to love," and as Rachel Carson instructed "It is not half so important to *know* as to *feel.*" We therefore need to come to the realization, as did Hermann Hesse's over-intellectual character, Narziss, in the wisdom of engaging in a life such as that of his sensualist character, Goldmund, whose home is the earth, not merely the idea of it. For it is only through the direct and purposeful exposure to the ways of water, first by entering into an intimate ecopsychological relationship with the element, and then by directly implementing that wisdom through engaging water via the myriad opportunities of landscape architecture, that one can learn about both the external and the internal frontiers. Achieving just such a balanced relationship was what Thoreau was really all about, describing his life's hydrophilic activities as being "nature looking into nature."

# Part I
# Plunging In: Precepts

*Water is simply the whole thing, isn't it?*
  —Oprah Winfrey, twenty-first century CE


*Earth is a living planet woven together with water.*
  —Bernard Fischesser et al., twenty-first century CE


*Should the world be designated a genre, its main stylistic device would no doubt be water.*
  —Joseph Brodsky, twentieth century CE


*All is born of water, all is sustained by water.*
  —Goethe, eighteenth century CE


*Water is the matrix of the world and of all its creations.*
  —Paracelsus, fifteenth century CE


*All things arise from the principle of water.*
  —Vitruvius, first century CE


*Everything is water, water is all.*
  —Thales, sixth century BCE

# Mainstreams:
# The Cultural History of Water

Water infiltrates and is intimately associated with human culture. This can be easily appreciated through a brief review of the history of how water has directly influenced civilization. The emphasis here is on water as an agent of postmodern connectivity. In other words, given that the tool through which modernism works is text, that through which postmodernism functions is *con*text. Water, then, can be regarded as the very element that binds us all together to the Earth and to the Cosmos in a collective history. In this respect, the physical hydrosphere of Vernadsky fuses with the cultural noosphere of Teilhard de Chardin, becoming one. "Every human being is a summary of this planetary adventure of which water is the prime element," Fischesser et al. matter-of-factly state. Or in other words, as Tom Robbins quipped, "It has been said that human beings were invented by water as a device for transporting itself from one place to another."

## Amphibious Beginnings

Since the late fifties, a band of macaques inhabiting a place called Hell's Valley near Nagano, Japan began to enter the water of the hot pools to escape the winter cold. It started with the young monkeys, and then eventually most of the adults started to tentatively dip themselves. Today, though most of the several hundred members of the tribe are content to soak quietly, several have taken to submerging themselves to walk along the bottom in search of nuts and other food. Some argue that such a process has occurred before. The Aquatic Ape Hypothesis posits that between eight and ten million years ago, in an area near modern-day Somalia, a group of hominids found themselves trapped on an island by rising sea levels after a long period of drought. Forced into a semi-aquatic existence, this splinter group of wading apes followed the waterways south, making their way down the Rift Valley to where they would eventually evolve into other species that would in turn become *Homo sapiens*. In 2002, the anthropology world was rocked by the so-called Toumai skull being discovered near the shores of ancient Lake Chad, possibly pushing back the origins of human evolution to seven million years ago. Lucy and her African brethren certainly did lead a littoral life, as borne witness by the many fossils of lake life found in close proximity to the skeletal remains. Indeed, biochemical anthropologists now go so far as to maintain that the development of our larger brains was the direct result of a marine, or lacustrine, based diet. And some believe that many of the other physical traits that separate us from all other terrestrial mammals, such as relative hairlessness and subcutaneous fat, follow from our amphibious past.

It even seems that "we" have always been deep immersers, not merely shallow-water waders. Bony outgrowths in the ear canals are today possessed only by lifelong divers. Not only do pharonic mummies display this trait, but so too do *Homo erectus* and *Homo neanderthalensis* fossils going back millions of years. And presence of hominid artifacts in southern Spain that are over a million years old suggest that some *H. erectus* took a shortcut and possibly swam across the Strait of Gibraltar rather than wandering all the way around.

## Water World

Geologists have now, after centuries of speculation and decades of scientific research, identified the probable location of the biblical flood of Gilgamesh, certainly one of the most significant events at the dawn of cultural history. Ten thousand years ago, the melting of the continental icecaps caused sea levels to rise, eventually leading to a breaching of the Bosphorus and flooding of the Black Sea, which at that time was a glacial meltwater lake around which people farmed. The massive diaspora from this inundated Eden gave birth to civilizations located on other freshwaters, this time riverine: the Nile and the fertile crescent of the Tigris and Euphrates basins nearby, and those of the Indus and Yellow Rivers to the east. This period of trans-migration and the instability that must have ensued can be glimpsed in the massive fortifications found at Jericho on the banks of the Jordan River, regarded by many as being one of the very first cities.

From the start, civilization is therefore saturated with water. Indeed the first word in the Persian dictionary is *ab*, for water, from which we get the word "abode," and the Persian word abadan, meaning civilized. And for ancient Mesopotamians, "Ea" was a god who resided in the "House of Water," the first two letters of which name forms "Earth."

The remains of other freshwater-based cultures have been recorded from many locations around the globe. The early Neolithic ruins of La Marmotta, now submerged under Lake Bracciano near Rome, are among the oldest of all settlements found in Europe. Lough Gur in Ireland is surrounded by numerous late Neolithic structures and contains islands referred to as crannogs, which, though looking completely natural today, are actually human-made defense structures that in the subsequent four thousand years have become colonized by plants and rooted to the lake bottom. Bronze Age communities once existed in the littoral zone of Lake Neuchatel in Switzerland and in the Biskupin wetlands in Poland, as did moated communities of the Yayoi Period throughout ancient Japan. More recently, the Aztecs in the central plateau of Mexico constructed a city beside and on top of Lake Texcoco that contained hundreds of canals, to an extent that greatly exceeded anything built subsequently in either Venice, Suzhou, or Amsterdam.

Cultural ecology has always been closely associated with hydrology. In ancient Egypt, one of the most important jobs was that of the pharaoh-appointed hydrologist who would enter the angled corridors of "nilometers" in order to measure the height of the river, and thereby

5

deduce about the expected time of the coming inundation. Romans left not only a wonderful legacy of roads scattered about their empire, but also an elaborate system of aqueducts and cisterns. Some of these latter structures, such as those at Masada, were later to be used against the Romans. Medieval castles and the regional sovereignty that they sustained frequently relied on the defensive use of water in moats. The quanats of the Middle East and irrigation systems of Bali are still some of the most impressive hydro-engineering feats the world has seen. Water has been used as a weapon of war, the Dutch piercing dikes to stop the French armies in 1692 and the Chinese blowing out levees to thwart the invading Japanese in 1938 (indeed, the latter event, with its loss of over a million lives, remains the largest death toll due to a single act of war). The population growth and physical development of many European cities was closely dependent upon progressive hydrologic planning. In the New World, the birth of the industrial revolution began in New England with the construction of canals and dams used for harnessing water power for the numerous textile and manufacturing mills. And it is these, through fostering a sense of economic self-reliance, which would go on to play an important role in America's eventual move toward formal independence from Britain.

### In and Out

Water, as well as being a direct requirement for human life, also serves as an important resource due to the non-human life that it harbors. The science of aquaculture and hydrophonics go back over a millennia and were integrally associated with the development and expansion of ancient societies in southeast Asia and also in the Mexico Valley where the "floating gardens" of Xochimilco are still harvested. Fishponds were integral to ancient Roman country estates. Today, the harvesting of freshwater fish is still an important source of sustenance in many regions of the world such as central Africa.

Water is also crucial for the production of beverages. Thousands have made the pilgrimage to the Guinness brewery along the banks of the Liffey in Dublin. The word, whiskey, comes from the Gaelic *uisce baetha*, meaning "water of life." Indeed, agricultural historians continue to debate whether the first grain crops were actually sown and harvested for beer or for bread.

Water has also played an integral role in cultural history as a receptacle for our waste. In addition to being wonderful imitators of Greek art, Romans also copied their sanitary engineering, as witnessed by the striking similarity between the communal latrines of the Roman port city of Ostia and those of Hellenic Corinth. Concerns about water and waste have always been an important design feature in architecture, such as the clay pipes found at Argos in Greece, or the open swill channels at Fountains Abbey in Yorkshire. And the development of many cities, such as London and New York, was closely related to the building of wastewater infrastructure.

## *Hydrophilia*

Water served as the focus for the birth of tourism. The hot springs at Lake Tiberias in Israel, like those at Bath, England and many other places, have drawn thousands of bathers annually since the days of the first Roman visitations. Healing spas have a rich sociological and architectural tradition in many places in Europe, North America, and Japan. Bodily immersion comes from a tradition with a long history. For example, the Christian practice of baptism in the Jordan River owes much to the nearby cult of the Essenes, who collected water from wadis along the shore of the Dead Sea. Through use of a network of elaborate channels, they shunted this water into massive cisterns several hundred feet deep, large enough to hold an entire year's supply for drinking and ritualized bathing.

Purification by water is an important element found in many religions. In ancient Greece, visitors to Delphi would have to perform the appropriate ablution to Apollo at the Castalian Fountain Spring prior to being granted access to the Oracle. The *mikvah*, or ritual bath, has always been a tradition for Orthodox Jews. For Christians, the ritual of baptism continues to be necessary for salvation: "Verily, verily I say unto thee, except a man be born of water and of the Spirit he cannot enter into the kingdom of God" (John 3:5). Every week, millions of Muslims ritually wash their feet before entering mosques the world over, remembering the message "From water we made all living things" (Qur'an XXI:30). For the Tal people of China, water is regarded as holy matter into which to immerse several times daily and in which to bathe the statues of the Buddha every New Year's Day. And during February 2001, in what was billed as the largest single gathering in human history—the *Kumbh*—tens of millions of Hindu bathers plunged into the Ganges to wash their souls: "Whatever sin is found in me, whatever wrong I may have done, if I have lied or falsely sworn, Waters remove it from me" (Rg Veda).

Water has functioned as a vehicle for worship, mystery, and myth due to it's pervasive haunting of "the human imagination," and its hovering "between the human and supernatural worlds," as Fischesser et al. state. Many creation myths are centered upon water. Tlaloc, the ancient Aztec god, was sacrificed to, the flowing blood of the human victims celebrating the hydrological cycle. The Cogui tribe in the Colombian forest, for example, regard creation as "water thinking." Vikings buried gold in bogs to appease the gods residing therein and Irish Celts offered golden treasures to Loughnashade, a small lake. Germanic tribes made sacrifices of slaves to *Nerthus*, a goddess of fertility, by plunging the victims into lakes or marshes. Ancient Gauls honored the source of the Seine River by casting wooden idols into the waters. And in Europe today, effigies of Death and Carnival are still thrown into rivers. For the Lozi people of Zambia, water is used as the basis of prayer, and hell is envisioned as a frightening place where there is no water. In Somnatha, a coastal town in India, pilgrims bathe a stature claimed to be 30,000 years old with water transported from the Ganges; as the statue is also washed by the tides, believers proclaim that even the ocean and moon join in adulation. In many Far Eastern traditions, the leaching of ink from prayers immersed in water and then released into nature aids resolution of the inscribed hopes. Thousands of holy wells dot the landscape of northern Europe, the emerging water venerated due to its contact with the underworld. Many of these wells peaked in popularity during the plague years of the Middle Ages. Even today, there are areas such as Derbyshire, England, where locals still remember and celebrate the life-saving properties of such springs with ceremonies in which the wells are decorated in thankfulness with flowers. In Greece, the *agiasma*, or "holy water," of wells is still a central feature in all churches that are built, not coincidentally, right atop primordial springs. And "diviners"—those who dowse for underground water—would search for the presence of divinity in such secret places.

There are thousands of lakes and rivers the world over that contain mysterious denizens. Some of these, such as the monster of Loch Ness in Scotland, are well known. Others, such as the phantom water nymphs of the Mummelsee in the Black Forest of Bavaria, or the riverside ogres of western Kenya, are only whispered about in hushed tones inside local taverns or tribal meeting houses.

People are drawn to water as locations where fiction fuses with history, nature with culture, all becoming confounded into myth. West Lake in China, Grassmere and Windermere in the English Lake District, the Sea of Galilee in Israel, Lake Dal in Kashmir, Lake Biwa in Japan, and Walden Pond in Massachusetts are visited by a combined total of over twenty-five million pilgrims annually. Though Galahad, Parsifal, and Bors might have had a tough time of it during their trials and tribulations in the original search for the Grail, today's visitors to Glastonbury, England, are aided in their quest by the presence of interpretive signs clearly pointing the way to the Chalice Well. Sherlock Holmes aficionados flock to Reichenbach Falls in Switzerland to see the location where the master detective and his archenemy Moriarity took their fateful plunge. In all these examples, who is to say what is really real, since culturally accumulated experience can often provide a legitimate substitute for authenticity.

### Water on the Brain

"We are able to think because our brains are afloat on water's buoyancy," wrote Theodor Schwenk. Witgenstein often drew the analogy between philosophical thinking and swimming: "Just as one's body has a natural tendency towards the surface and one has to make an exertion to get to the bottom—so it is with thinking." That said, water has inspired more than its fair share of pseudo-scientists and wet wachos believing in its mysterious energetic properties that become manifest through flow forms, ripple tanks, frozen crystals, and the like.

In the sixth century BCE, the Greek natural philosopher Thales, who would later be an inspiration for Thoreau, considered all the material of the world to be derived from a single worthy element—water. Lao Tzu in turn believed that "the highest good is like water," for it was water that gave life, did not strive, went to all places, and so was therefore "like the Tao." Such a view was later adapted by Hegel in his *Philosophy of Nature*: "Water is the element of selfless contrast, it passively exists for others . . . its fate is to be something not yet specialized . . . and therefore it soon came to be called 'the Mother of all that is special.'" During the Middle Ages in Europe, water served as a major focus for alchemical inquiry, continuing into the seventeenth century when van Helmont reached the same conclusion as Thales: that there was but one true element, and that was water. Indeed, in many respects, the history of science is the history of water. "Physics" and "physiology" both have their root in the Hellenic Greek *physis*, referring to the watery substance that connected all lives with the natural world.

The close linking of water with philosophy is no accident, since water as an element reveals and reflects, the word "reverie" deriving from the French term for river. Poets in particular have long used water imagery as a metaphor for life: lakes represent melancholy, rivers represent flowing time, depths of water parallel depths of thought, nature's liquid is the embodiment of humankind's collective tears, and the reflective surface of ponds serves as a mirror in which to learn about ourselves. In dreams, water invokes myth and memory, representing both death and life, serving to cleanse the earth, and, due to its reflective properties, doubling the seen world by signifying our celestial aspirations as it seems to grasp the very sky filled with the purest of our heavenly thoughts.

### Wet Musings

No element has so inspired writing as has water. Poets "brim over" with feelings, and the worst fate a writer can have is to "dry up." Water has frequently played an important role in literature as in, for example, George Eliot's *The Mill and the Floss*, which deals with river management and interfamilial relationships in the English countryside of the nineteenth century. More recently, there is no understating the importance of Frank Herbert's *Dune*. In addition to being one of the most popular books ever written, its appearance in the late sixties coincident with the first pictures of the Earth from space, as well as Herbert's message of a planetary, Gaiaesque water homeostasis, were important factors leading to the first Earth Day meetings in Stockholm several years later. Rivers have often been employed in fiction to symbolize journeys of transformation, as in, for example, the River Liffey in James Joyce's *Finnegan's Wake* or the Mississippi in Mark Twain's *Huckleberry Finn*.

Water continues to be a focus for popular literature such as Norman Maclean's *A River Runs Through It*, David Duncan's *The River Why*, or

Andrea Barrett's *The Forms of Water*. Modern essayists have often focused on water. Edward Abbey, when not writing about deserts, was next most enamoured by water. In his *Down the River* collection, perhaps the most memorable piece describes a riverside chat with the ghost of Thoreau. Annie Dillard, one of the most well regarded nature-culture writers, wrote the award-winning *The Pilgrim at Tinker Creek*, concerning a modern-day experiment in Thoreauvian inhabitation beside a body of water.

Water and literature also conjoin in the whole mythologized world of fishing. Richard Brantigan's *Trout Fishing in America*, in which water and fishing serve as metaphors for self-exploration, was one of the most popular books of the sixties. *The Compleat Angler* by Izzak Walton is, next to the Bible, the most published book in the English language. Here fishing becomes a springboard into religious salvation in a sort of aquatic *Pilgrim's Progress*.

### Reel Nature

Since the first theatrical screening of any film, that of *The Waterer and the Watered*, in Paris in 1895, cinema has often returned to water as a subject such as *Rain*, a 1929 film about a downpour in Amsterdam. Some of our very best films are based on water management, whether on the scale of dealing with the machinations of corrupt government officials and gangsters in 1930s Los Angeles, in Roman Polanski's *Chinatown*, or on the more intimate scale of a small, nineteenth-century French village in *Jean de Florette* and *Manon des Sources*.

Water in film is often used as a vehicle for human change. This can occur for an entire society, as in Peter Weir's *The Last Wave*, which deals with Aboriginal Dream Time and the arrival of a biblical-like cleansing flood, or in Fernando Solanas's *The Voyage*, in which a flooded, sewage-filled city mirrors the corrupt political pollution. Water-inspired change can also occur for individuals, as in the *The Rainmaker*, in which the arrival of water and the character Starbuck invoke a shift in personal development for the female protagonist, or in *Dark Eyes* in which the protagonist's visit to a spa coincides with a romantic infatuation. And certainly, Janet Leigh's watery end in *Psycho* is justifiably famous.

Even in films not ostensibly about water, the element can still feature prominently in the transformation of the characters. For a scuba-diving Dustin Hoffman, water in a swimming pool provides solace and escape from an adult world gone mad in *The Graduate*. In contrast, for Martin Sheen in Francis Ford Coppola's *Apocalypse Now*, water functions in quite the opposite way, the river along which he is being transported serving as a thin blue line into his own heart of darkness.

And finally, water is often used as a symbol for dramatic effect, perhaps nowhere more memorably than in Carol Reed's *The Third Man*. Here, in what has been called the most famous entrance in film history—the shot across the rain-sodden streets to where Orson Wells is lurking in the shadows—and then later, when Wells's character, the despicable Harry Lyme, meets his just end in the CSOs or combined sewer outfalls of postwar Vienna, dirty water signifies moral pollution.

### Artistic Inspirations

Handel's *Water Music*, Brahms's *Rain Sonata*, Schubert's *Trout Quintet*, Debussy's *La Mer*, and Strauss's *The Blue Danube* are just five of the well known and distinguished classical compositions that have been inspired by water. Today, New Age musicians continue to focus on water in the production of what might be called "healing music." Albums with titles such as *Dream Streams*, *Healing Waters*, and *Peaceful Ponds* are reported to be frequently listened to by no less a personage than the Dalai Lama himself.

The Hudson River School of landscape painting developed during the mid-nineteenth century. Fleeing the increasing development of New York City, a loose group of artists including Frederic Church sought inspiration in the pastoral headwaters of the Hudson Valley. Water was frequently used as a metaphor upon which to reflect about the divinity considered present in the bucolic countryside. The movement reached its apogee in religious symbolism with the sublime landscapes of Thomas Cole, such as the *Voyages of Life* series and, especially, *The Holy Vessel*.

The best known group of painters to be drawn to water were the late-nineteenth-century Impressionists. For these artists it was the attraction of water itself and the reflections of light upon it that led to such masterpieces as Cezanne's *Chalet on the Lake* or Renoir's scenes of boating along the Seine. Indeed, it could be argued that the best known image we have anywhere of riparian recreation is Searat's *Afternoon in the Park*. And of course, probably most famous of all, are Monet's grandiose paintings of water lilies from his garden at Giverny.

The Group of Seven painters were productive in eastern Canada in the years between the wars. Abandoning concepts of the English pastoral tradition, and being dismissive of the transcendental religiosity of the Hudson River School, this group of artists sought to capture their young and developing nation on canvas. To effect this, they found inspiration in the rugged harshness of the Canadian Shield landscape north of Lake Superior. Their paintings of lakes and rivers, often through filtered views of trees, are among the very best images ever produced showing water in the undomesticated wilderness. A. Y. Jackson's *The Red Maple* led to establishment of the maple leaf on the Canadian flag in the sixties, and Tom Thompson's *The Jack Pine* remains the single archetypal Canadian painting.

Numerous paintings have also featured humans closely associated with water, including John William Waterhouse's *A Mermaid*; *Ulysses and the Sirens*; *Hylas and the Nymphs* and *The Lady of Shallott*. Death and immersion have long been a popular theme, as witnessed by John Everett Millais's *Ophelia* and Jacques-Louis David's *Marat Mort*.

Dance has also been inspired by water. For the performance of the first ballet, 1581's *Le Ballet Comique*, a giant rolling water fountain was employed. And Isadora Duncan often would dance atop a blue-green carpet to remind her of water.

### Waterscape Architecture

There is a long tradition of using water to frame sculpture and buildings. The Canopus is easily the most recognizable feature of Hadrian's Villa estate. The courtyards at the Moorish palace at Alhambra in Granada are characterized by the imaginative use of water. These runnels are a feature of Persian, Mughal and Islamic gardens and go back to use in early Bronze Age cities such as Ugarit. Water has also often been used to create reflecting pools like those at the Palace of Versailles, the Fahtepur Sikri temple in India or at the Christian Science plaza in Boston. The latter is truly an engineering marvel in the way the recirculating water pours evenly over the block-long lip of the upper pool. The architecture of pools specifically created for swimming and bathing has always been an important element in civic infrastructure.

Probably no city celebrates water more beautifully than does Rome with its many fountains, such as those by master architect Bernini. Fountains have of course played a very important role in the history of European formal gardens; an outstanding example is the Villa d'Este outside Rome. On a more intimate scale, fountains such at those in front of the Harvard Science Center or at several locations in downtown Portland, Oregon are frequently crowded with hordes of children on hot summer days, and have been heralded as being among the most significant developments of modern landscape architecture. And in Las Vegas, water, just like so much else there, is celebrated in exuberance, with the world's largest and most elaborate fountain as well as several grandiose water fantasy parks.

*Not every water healed. The Red Fountain of Ethiopia brought delirium;*
*the dark waters of Avernus, like those of Asphaltites which Pilate himself*
*looked upon, perhaps touched, perhaps tasted, swallowed both leaves and*
*birds. In Macedonia, the Lake of Insanity became salt and bitter three times*
*a day, and crawled with white serpents twenty cubits long. The waters of the*
*Styx brought death at once; those of Leontium killed after two days. At*
*Tempe in Thessaly, out of a bank overhung with wild carob and purple flow-*
*ers, a small stream trickled that corroded copper and ate the flesh of men.*
> —Ann Wroe, *Pilate: The Biography of an Invented Man,*
> 2000


*Matters have now reached a point where the element upon which life and*
*health depend must itself find healing if human life on earth is to continue.*
> —Theodor Schwenk, *Water, The Element of Life,* 1989


*You cannot fill the Aral Sea with tears.*
> —Uzbeck poet Mukhammed Salilch *in* Philip Ball,
> *H₂O: A Biography of Water,* 1999

# Undercurrents:
# Sources of Aquatic Anxiety

### *The Sins of the World*

Lakes serve as convenient barometers for changes
occurring within their watersheds and airsheds.
Eventually everything we do upon the land and inject
into the atmosphere that then rains down upon the
land, will find its way to the bottom of lakes.
"Paleolimnology," from "paleo," meaning ancient, and
"limnology," meaning the study of freshwaters, is the
field of investigation where scientists take cores of
sediments from the bottom of lakes and, by dating the
layers like tree rings, inform us about the historical
anthropogenic and natural alterations in the environment surrounding and supporting those lakes.

We are generally accepted as being on a collision
course with the natural world: "Human activities inflict
harsh and often irreversible damage on the environment and on critical resources. If not checked, many of
our current practices put at risk the future we wish for
human society." This statement, not from some tree-
hugging, left-wing environmental lobby, but rather

17

comprising a warning letter signed by more than half of all living Nobel Prize laureates in 1992, underscores the magnitude and severity of the problem.

So, no matter where on the planet, there is often much for paleolimnologists to measure in the bottom of lake sediments. But what if, in addition to gauging the effects of physical pollution upon the landscape, paleolimnologists could also somehow detect the litany of moral pollution that accompanies those physical alterations?

Thoreau would certainly have thought there was nothing strange about entertaining such a question. For example, after conducting one of the very first bathymetric surveys ever undertaken of a lake, he remarked that "what I have observed of the pond is no less true in ethics." Therefore, plumbing the depths of Walden Pond, Thoreau explored the depths of his own soul: "My nature may be as still as this water," he wrote, "but it is not so pure." This is a remarkable idea, and also a truly frightening one—namely that water could somehow register a memory of every act that we have ever inflicted upon it. Of course this is fiction, but consider for just a moment, what if it were true.

Contrary to the erroneous supposition by many, water is a not a renewable resource. There is neither more nor less water around now then there ever has been or ever will be; no more can therefore be created. So every molecule of every droplet of water in existence today has always been there, recording all our acts upon the globe. Imagine taking one of Thoreau's ethical cores from the bottom of a lake, the sediments there integrating all that that lake has witnessed throughout its existence. Imagine too that like Pilate, untold millions have washed their hands free of crimes, and that water then carried those sins away to the bottoms of lakes, where they reside in some sort of museum of shame. One wonders what might that look like?

### For the Time Being

| | |
|---|---|
| —1,360,000,000,000,000,000 | cubic meters = the global supply of water (the most abundant element on the face of the Earth) |
| —5,400,000,000 | gallons = the amount of bottled water sold worldwide in 2001 (more than double the amount sold in 1991) |

18

—4,800,000,000          cubic meters = the amount of saltwater converted to freshwater annually by desalinization plants

—4,200,000,000          people = the number of individuals living within one kilometer of surface water

—1,500,000,000          liters = the daily supply of water used by New York City (which is about equal to the flow of the Hudson River)

—1,200,000,000          people = the population in the developing world that do not have access to safe and reliable drinking water

—300,000,000          liters = the amount of used motor oil dumped into Canadian waters each year (which is equal to seven times the spill amount from the Exon Valdez)

—235,000,000          hectares = amount of land currently being irrigated

—100,000,000          people = the worldwide population drinking bottled water from two large French companies

—20,000,000          tons = the amount of fertilizers dumped annually on U.S. croplands, a portion of which is transported into waterways

—20,000,000          people = the global incidence of waterborne schistosomiasis

—20,000,000          pounds = the daily contaminant load released by Detroit into Lake Erie in its 1.6 billion gallon wastewater discharge

| | |
|---|---|
| —15,000,000 | people = the number of deaths per year due to waterborne diseases (4 million due to diarrhea alone) |
| —5,000,000 | tons = the amount of oil expelled into water from the world's refineries each year |
| —2,500,000 | cases = the global incidence of infectious hepatitis from consumption of contaminated shellfish |
| —2,000,000 | hectares = the extent of agricultural land lost for production per year due to poor drainage practice |
| —2,000,000 | tons = the daily amount of untreated human excrement discharged into rivers and groundwater globally |
| —1,000,000 | hectares = the amount of cropland lost to production annually due to salinization from irrigation |
| —625,000 | tons = the amount of pesticides and herbicides dumped on farm fields annually in the U.S., much of which runs off into receiving waters |
| —600,000 | cases = the global incidence of waterborne cholera in 1992 |
| —500,000 | hectares = the amount of cropland in 15 countries that is irrigated with municipal wastewater |
| —291,000 | gallons = the amount of water needed to grow food for a low-meat diet for 1 person for a year |
| —250,000 | people = the number killed in 1975 when a dam collapsed in China |

—200,000 square miles = the extent of wetlands that have been drained in the United States through channelization

—200,000 tons = the amount of oil deliberately released by Iraq to hinder American operations in the first Persian Gulf War

—100,000 gallons = the amount of water needed to produce one automobile

—75,000 dams = the number in the U.S.

—46,000 infractions = the number of times U.S. drinking water systems have violated purity levels

—38,000 dams = the number of major hydrological development projects in the world with dams exceeding 6 m in height

—18,770 locations = the number of impaired water sites listed by the EPA

—15,000 tons = the amount of sodium chloride dumped into the Rhine every day from a single potash mine in Alsace from 1930 to 1980

—8,000 cubic meters = the annual water supply available per person

—7,500 plants = the number of desalinization plants worldwide

—6,350 species = the number of aquatic species that are either already extinct, critically imperiled, or considered vulnerable of the total 20,481 that exist in the U.S.

—5,320     liters = the amount of water needed to process a meal of a quarter-pound hamburger, French fries, and a soft drink

—5,000     people = number that could be sustained by the amount of water used by each golf course

—4,153     incidence = the number of beach closings in the United States in 1979 due to health risks

—2,800     BCE = the year archeologists believe the world's first drainage irrigation systems were built in Mesopotamia

—2,740     liters/person/day = the level below which nations are designated as being "water scarce"

—2,600     BCE = the year archeologists believe the world's first dam was built in Egypt

—2,500     liters = the amount of water required to produce 1 pound of rice, the crop which comprises half the daily diet for 1 out of every 3 people on the planet

—2025     year = the year in which it is estimated that most Asian countries will have severe water problems

—2,000     miles = the extent of streams in Pennsylvania devoid of most aquatic life due to acid mine discharge

—1,870     miles = the extent of main-stream levees constructed along the Mississippi to isolate the river from its floodplain

—1968     year = the year in which Cleveland's Cuyahuga River caught fire twice

—1,500     gallons = the amount of water needed to produce a cotton dress

—1,279     liters = the average daily use of water by a U.S. family of 4 (in contrast to that of 19 liters for most other countries)

—700     kilograms = the amount of water needed to produce a single gram of paper

—605     liters/person = the per capita daily supply of water from public systems in the U.S.

—600     kilometers = the length of the reservoir to be formed behind the Three Gorges Dam in China, the world's largest

—470     trees/mile = the extent of removal of fish habitat over 27 years in Oregon waterways by the U.S. Army Corps

—400     kilometers = the distance over which Los Angeles must transport its water

—300     meters = the increase in depth of wells in Tucson, Arizona needed in order to reach water

—300     times larger = the American daily consumption rate of water compared to the Ghanaian rate

—240     kilometers = extent of the 8,000 km U.S. shoreline of the Great Lakes that is unfit for swimming due to health concerns

—226 days = the length of time the lower part of China's Yellow River ran dry in 1997 and failed to reach the ocean

—200 meters = the height of the Colorado River impounded as Lake Mead behind the Hoover Dam

—175 liters = the daily consumption rate of water for a resident in Canada for domestic use (compared to 10 liters per person for someone living in northwest Africa; i.e. the latter amount is the same as that used by a Canadian to flush the toilet once)

—190 liters = the amount of water used per cycle in a washing machine

—120 meters = the drop in level of the great Artesian Basin in Australia over the last eight years due to withdrawal

—106 extinct salmon breeding stocks = the number of the 1,000 that once existed in the Pacific Northwest

—100 gallons = the daily per capita rate of wastewater production in U.S. cities

—100 meters = the rate at which Lake Chad's shoreline is receding per year

—95 percent = the maceration-mortality rate of salmon smolt as they go through dam turbines

—83 kilometers = the length that the Rhine has been shortened due to channelization for navigation

—80      percent = the proportion of rivers in China that are too degraded to support fish

—80      species = the number of upstream fish species expected to be extirpated following completion of the Three Gorges Dam in China

—75      percent = the proportion of the global freshwater supply used by agriculture

—70      percent = the proportion of Israel's water consumption that originates in regions occupied or annexed since 1948

—70      percent = the average proportional rise in bilharzia incidence in human populations following dam construction

—70      percent = the proportion of Beijing's total water demand which is in excess of its available supply

—70      percent = the water content of humans if rendered down

—64      percent = the proportional increase in per capita withdrawal of water since 1900

—60      percent = the proportion of water intended for irrigation that never reaches the croplands due to leaky pipes, evapotranspiration, and unlined canals

—60    percent = the proportional decrease in the volume of the Aral Sea since 1960 due to water diversions

—56    kilometers = the length of the Kisimmee River in central Florida lost due to channelization for flood control

—50    percent = the proportion of the global wetlands thought to have been destroyed in the last century alone

—45    percent = the projected decrease in river flows in western Australia due to climate warming

—42    feet = the drop in water levels in Mono Lake due to diversion of feeder streams by Los Angeles

—40    times = the number of times that the planet could be encircled by the kilometers of pipeline and aqueducts constructed in the U.S. and Canada

—40    percent = the proportion of lake acreage in the U.S. that is classified as being polluted

—33    percent = the proportion of all Africans that qualify as living under conditions of water scarcity

—33    feet = the depth that the historic center of Mexico City has sunk over the last century due to groundwater withdrawl

—33      percent = the proportion of a woman's total daily caloric intake expended in carrying water in many parts of the world

—33      percent = the proportion of towns in northwest China whose drinking water supply has gone dry

—31      percent = the proportion of the flow of the Danube that is extracted for human use

—30      percent = the proportion of agricultural water supply in Israel that is treated wastewater

—30      tons = the amount of poisons (herbicides, pesticides, mercury, dyes, and heavy metals) released into the Rhine at Basal in 1986 due to the collapse of a retaining wall

—30      percent = the proportion of river miles in the U.S. that are classified as being polluted

—30      kilometers = the distance that the Aral Sea has receded from the former port town of Mujnak

—25      times greater = the rate at which the Ogallata Aquifer in the western U.S. is being pumped relative to its rate of refill

—24      percent = the depleted proportion of the Ogallala Aquifer in Texas

—20      species = the number of freshwater mussel species that have become extinct in the U.S. over the last century

—18     percent = the proportion of the world's freshwater fish species that face a moderate to high risk of extinction

—16     percent = the proportional decrease in output of the world's rivers due to climate warming

—16     percent = the proportion of the global population that lacks the convenience of a home faucet

—15     percent = the proportion by which Israel's water use annually exceeds its supply

—13     percent = the proportional flow of the Jordan River in 1998 compared to what it was in 1953

—10     gallons = the amount of water needed to produce one copy of a book such as this one

—10     percent = the proportion of worldwide irrigated land that suffers from yield-suppressing salt buildup

—10     meters = the drop in level of the Dead Sea over the last century due to reduced inflows from the Jordan River

—9     percent = the proportion of the population of Mexico City that uses 75% of the city's total water supply

—8     feet = the height of discolored detergent foam that would accumulate at the base of Niagara Falls during the 1960s before phosphate additives were banned

—8        percent = the proportion of world's water supply located in China which has 22% of the world's population

—8        minutes = frequency at which a child dies from waterborne diseases

—8        meters = the drop in the water table in Karnataka State in India between 1946 and 1986

—8        dollars = the price which customers in Boston in 2002 were willing to pay for a single bottle of water at an expensive resturant

—8        times greater = the relative difference between the rate of our water consumption compared to that of our grandparents

—3        percent = the proportion of global waters that are freshwaters

—3        times = the rate of increase in global water use since 1950

—3        percent = the proportion of the daily water use by families in the U.S. that is used for drinking or cooking

—3        meters/year = the rate at which the Nile Delta is eroding due to sediment being trapped behind the Aswan Dam

—2.5       percent = the proportion of  riparian forest remaining in Arizona

—2        percent = the proportion of the flow of the Orcontes River remaining after it passes through Syria

| | |
|---|---|
| —2 | times = the difference in the rate of increase in water usage compared to that of the human population between 1900 and 1990 |
| —2 | percent = the proportion of sewage treated by Santiago before it is discharged directly onto the city streets and into the rivers |
| —1.8 | percent = the proportion of the 3,230,000 stream miles in the U.S that are deemed "pristine" |
| —1 | meter = the annual drop in the water table over an area of millions of hectares in northern China |
| —1 | lake = the number of the great African lakes that are free of schistosomiasis and therefore safe for swimming |
| —1 | town of 10,000 people = the equivalent supply of water used by a single golf course |
| —1 | person (Victor Hugo Daza) = individuals assassinated at a Bolivian protest against water privatization by foreign companies |
| —0.917 | percent = the proportion of Japan's major rivers that remain undamed |
| —0.9 | percent = the proportion of the global water supply that exists in inland surface waters |

—0.0025 percent = the per capita proportion of the total water supply of the Gaza Strip used by Palestinians (in contrast to the per capita proportion of 2 percent used by Israeli settlers; i.e. 800 times greater)

—0.00000000000000000000 cubic meters = the amount of water created anew every year; i.e. water is not a renewable resource

### Drowning by Numbers

Numbers numb. Statistic stun. What are we to make of the frightening observation that acidification has killed 2 thousand miles of streams in Pennsylvania? Or that detergent foam once piled up to a height of 8 feet at the base of Niagara Falls? And how do we really comprehend the enormity of 20 million pounds of contaminants being released daily from Detroit into Lake Erie? Or that over a billion people on the planet do not have access to safe drinking water?

We have become inured to the constant barrage of facts, factoids, and faxes that cry out for attention at the margins of our mostly comfortable, modern, Western lives. "Details don't matter," we fool ourselves into thinking. "Just give me the big picture," we intone, as if somehow we imagine ourselves being both strong and smart enough to receive and process such news. "NUMBERS CAUSE MIND TO GO SLACK, the *Hartford Courant* says," begins one section in *For the Time Being*, where Annie Dillard warns about the dangers of succumbing to this cultural neurasthenia: "But our minds must not go slack," she admonishes. "How can we see straight if or minds go slack? We agree that we want to think straight." How does this "thinking straight" affect the need for the special "water consciousness" that Theodor Schwenk wrote about again and again, which might be conceived of as a "fluidity of mind"?

And what about the inevitable inaccuracies no doubt present in the long litany of numbers and statistics cited above? Do we really know for sure that it is 300 million liters of used motor oil that are dumped into Canadian waters each year? What if it were only 250 million? Would that be reassuring? And what if the per capita proportion of the water supply in the Gaza Strip allotted to the Palestinians was really 0.025 percent rather than 0.0025 percent compared to the 2 percent used by Israelis? Would we feel all that much better? Would that somehow reduce our collective guilt?

And what about the occult numbers that are buried in those provided? How many individual lives are represented in the statistic that 20 freshwater mussel species have become extinct in the United States over the last century? Or that 106 salmon breeding stocks have disappeared from the Pacific Northwest? Are the numbers 20 or 106—both certainly easy to wrap our minds around—more or less bothersome than the untold millions of those creatures that are now gone forever? Were they even all that important in the first place, given that we are still here to write and read about them? If we think so, and now that they are gone, are they important enough to remember for future generations? In James Cowan's *A Mapmaker's Dream*, a medieval cartographer struggles to find ways in which to map the presence of vanished races that his traveling sources have told him once existed. How do we map all those lost aquatic lives upon the landscapes of our memories?

And what about the harsh future realities implied in the numbers? How long can Lake Chad's shoreline keep receding at a rate of a 100 meters a year? Will it "win" the race to dry up before the Aral Sea does? How about Mexico City? At a subsidence rate of 36 feet a century, when will it bottom out? Are there other cities in the competition? It is human nature to imagine trends being linear. What happens if they are not—if they are instead exponential? In that case, we are forced to realize that we have even less time to right the wrongs. Does this knowledge inspire action or promote apathy?

And finally, what about the potential synergistic interactions hidden behind those numbers? If 33 percent of all Africans now live under conditions of water scarcity, how long can they continue like that when half of the population, the women, expend 33 percent of their daily caloric intake carrying what little water that they do manage to find? It is reassuring to conceptually treat facts and figures as if they were somehow independent, but if they are not, and if the interactions are multiplicative rather than additive, certainly the situation becomes much worse than we imagine it, does it not?

"The pain of water is infinite," wrote Gaston Bachalard. In the end, his favorite poet, Alponse de Lamartine, might have put it best when he said that "water is the sad element." Because it weeps with everyone, water serves as a receptacle for the world's tears. To drink from a bubbling spring, therefore, is to ingest history's sins realized in liquid form.

*The problem of rescuing water from death must therefore be solved inside ourselves before we can solve it in the external world. When we have transformed the inner scene, the outer one can be restored to order.*
    —Theodor Schwenk, *Water, the Element of Life*, 1989


*By seeing, hearing, touching, and tasting—by ingesting—we become the world within which we are.*
        —Laura Sewall, *Sight and Sensibility: The Ecopsychology of Perception*, 1999


*Once you start spending time in the wet, glimpsing eternity in glides, or being lifted above the earth by joys caused by nothing but water's flow, you tend to stop fretting.*
    —David James Duncan, *My Story as Told by Water*, 2001

# Head Waters, Water Balms:
## Liquid Solace

### *Lives of Quiet Desperation*

The domestication of our modern lives has left us sensually crippled. We have lost intimate contact with the natural world and have begun, collectively, to suffer from a deep pathological illness whose symptoms often become manifest as rampant materialism in a desperate attempt to fill in the grand lacunae that characterize our existence. This ecocidal behavior is due in part to the deadening silence, or apatheia, that many feel within, because of our inability to deal with the emptiness we see outside us. Depressed, we focus on changing the superficial, buying new trinkets: "consume, obey, be silent, die," as one bumper sticker heralds with black humor. Meanwhile, we are blind to the profound psychological transformations needed to rescue ourselves. Rising to the imagined panacea of material abundance, we struggle all the more yet, ultimately defeated, are cast down into the darkness, far deeper than before.

With assured confidence, we have used every technological means at our disposal to do the very best to remove ourselves from the world. Our lives are lived in boxes designed to be hermetically sealed from the vagaries of weather as much as possible: the dash from the air-conditioned house to the air-conditioned car in the garage whose door we open and close electronically, then the drive to the air-conditioned office at work where we park underground. It is now almost possible to live a life completely outside the environment, never getting hot, cold, or, most dreaded of all, wet. Cut off from the seasons, we have become cut off from the life spark of the world and, as a result, have become sensually numbed. Neurasthenia, it has been said, may be regarded as our most successful collective accomplishment of the last several centuries. How have we gotten this way? How do we go about attempting to heal the "world wound"?

"In wildness is the preservation of the world," Thoreau instructed. "Wildness," he wrote, not, as is often misquoted, "wilderness." The distinction is an important one in that it brings what many erroneously perceive as being a problem "out there" to being one "in here." For more than anything else, it is the alienation from our own inner wildness that resides at the root of our current disharmonious relationship with the natural world. As Thoreau wrote "It is vain to dream of a wildness distant from ourselves. There is none such."

We need, therefore, to rekindle our sense of a love of wildness. By loving that which is wild within, we can then love that which is wild without. As Laura Sewall remarks: "Below the ecological crisis lies a deeper crisis of love, that our love has left the world, that the world is loveless." We need to accomplish what the poet Robinson Jeffers referred to as "falling in love outwards."

Deep problems necessitate deep solutions. The discipline of "deep ecology" developed over the last several decades as a system of thought and action that doesn't merely treat symptoms of environmental aliments, but instead questions fundamental premises underlying the current malaise. Recently there has emerged a cadre of concerned deep ecologists who refer to themselves as ecopsychologists.

The working precept of ecopsychology is based on the supposition that it is impossible to have well persons residing on a sick planet. In the light of this fact, therapists must begin to recognize that there is never a sharp divide between the inner and outer worlds, but instead a continuum existing from the scale of the planet to that of the person. In this sense, the world can be regarded as an extension of ourselves, with a synergistic interplay between the two. Ecopsychology, then, concerns itself with exploring the motivations, yearnings, needs, and ideals that shape and structure our lives within the environment, focusing on strengthening or even reawakening the reciprocal relationship.

Reciprocity here refers to the concept that through us, acting as its sense organs in the noosphere, the world in a way becomes able to perceive itself. And water, some believe, may play an important role in fostering such a relationship. Thoreau, for example, adapting the Narcissus myth, referred to lakes as being "the earth's eye, looking into which the beholder measures the depth of his own nature." Paul Claudel considered water to be "the gaze of the earth, its instrument for looking at time." And for Thomas Faber, water is also the earth's instrument for looking back at us, as well as enabling us to regard one another.

Given the supposition that changing the planet must come from within, is it possible to improve the state of the internal landscape by prescribing nature for healing, the hope being that with an improved personal health, one can then reciprocate and work toward restoring planetary health? Such ecotheraputic healing comes about through fostering and developing a healthy interaction with the earth, not just by improving the psyche, but by dealing with the total mind-body-spirit relationship. By engaging in such a practice, we begin to turn inside-out, in what David Abram believes will loosen "the psyche from its confinement within a strictly human sphere, freeing sentience to return to the sensible world that contains us. Intelligence is no longer ours alone but is a property of the earth; we are part of it, *immersed in its depths*" (my italics). Depth is "an impression on the body," says Laura Sewall. Recognizing that we are really *in* the Earth, not merely on it, and that through birth, we don't enter into the world, but come *out* of it, goes far toward helping us to remember our physical attachment to the world.

### The Way of the Flesh

James Hillman considers the soul as being that "which deepens events into experience," and from Blake we know that it is the senses that function as gateways to our souls. Given the deadening of senses that is at the core of the environmental crisis, reawakening those senses must be the first step toward re-engaging with the world.

There is great danger in overintellectualizing this process. We must get out of our heads and learn to open ourselves up by becoming visceral. This is why David Cumes instructs "It is better to experience the learning, than to learn the experience." We can then translate our inner beings into our physical bodies, and thereby move toward inner peace. Arne Naess, the founder of the deep ecology movement, stated that "I'm not so interested in ethics and morals. I'm interested in how we experience the world." The challenge, then, is to move from speculation to revelation by trading mental knowledge for the wisdom of direct experience.

A single event at a recent conference in Bergen, Norway, at which the International Water History Association was established, suggests, however, that the shift from learning to experiencing water will not come easy, even to those whose professional lives are intellectually immersed in the substance. As the boat moved along the painfully beautiful, cloud-draped fjords, almost all delegates were content to sit comfortably inside talking about water rather than being outside in the cold and damp wind, feeling the spray upon their skin; in other words, experiencing the very element that had drawn them from around the world to this particular location in the first place.

Transcending our minds, we must recognize that our bodies are the most concrete example of the natural world within our lives. The secret is to indulge in a phenomenological relationship with the world through direct experience mediated by the body in which we learn the texture, rhythm, and tastes of the physical world about us. In other words, we need to empower our eyes, skin, tongues, ears, and nostrils, and thereby awaken our bodies to truly experience the aliveness of this world. "By touching, tasting, sniffing, and looking again and again, we become tied to the earth, enlarged and wedded by the gift of having a body, of having eyes and ears with which to read the signs, with which to wrap ourselves within the great soul of the world," Laura Sewall states. "For," as David Abram similarly reflects, "it is only at the scale of our direct, sensory interactions with the land around us that we can appropriately notice and respond to the immediate needs of the living world."

Today, we live in a passionless world in which most people have no special desires, only preferences. Ours is a society of sensual eunuchs, impotent to the callings of the wildness within and as a result, the pull of that which resides outside. We desperately need to invigorate our lives through a collective Viagra-like dose of intense feeling provocation. We have to rediscover our bodies and how they interact with the world. "Care for the planet," Laura Sewall says, "follows true desire and passion, which naturally flow from seeing, touching, and tasting the beauty of our world. Like any love affair, loving the planet comes through the senses."

This is what William Calahan refers to as "groundedness," being the "dynamic state of the person that includes the sense of confidence, pleasure, and wonder resulting from progressively deepening contact with the wild." To Howard Clinebell, this is the process of "ecobonding"; that is, "discovering, befriending, and intentionally developing one's profound rootedness in the biosphere"—in other words: "biophilia." Certainly Thoreau's immediate descendant, John Muir, knew this well. He wrote that "the rivers flow not past us, but through us, thrilling, tingling, vibrating every fiber and all of the substance of our bodies, making them glide and sing."

### Feminine Fluidity

The principal tenet of "ecofeminism" is that the despoiling of the Earth and subjugation of women are both part of the same act. Mary Gomes and Allen Kanner discuss how this can be related to water in their allegory of an old Grail story called "The Rape of the Well-Maidens." Here, after the king has forced himself upon one of the guardians of the life-giving wells, the remaining maidens retire with their magic, causing the wells to dry up and the regenerative properties of the land to wither. And, in a modern and humorous Lysastrata-esque twist to the story, a group of women in an eastern Turkish village recently refused to have sex with their husbands until the men finally cleaned the water source and fixed the town well.

Many have recognized that women and water are intimately connected. Indeed, at a recent conference entitled "The Bonds Between Women and Water," papers were presented dealing with such topics as the role of women in transporting water in developing nations, the mythology of mermaids, the parallel life-giving natures of women and water, and the dampness of the act of childbirth itself. For example, water rituals such as *Aadiperukku* in India or *Kalegeros* in Greece, which are associated with fertility, are still openly practiced by country women residing there.

Almost certainly, though, it is the act of menstruation that more than anything else tightly couples women with water. As Lorraine Anderson has written "Our blood is a remnant of the great salty ocean in us. It flows with the tides, subject to the moon's pull." No one has explored this interlinking of the feminine and the aquatic with greater depth and insight than ecopsychologist Laura Sewall. Residing on an incredibly dynamic estuary on the coast of Maine, she exalts in the pull of water on her life as she envisions herself part of a vast tidal pulse: "Wet, yielding, and thriving, the grasses grow thick and luscious green with the flood of salt water. I too yield, the tide pulsing in me over many days and weeks, my blood flowing with the big moon tide. I watch the sea become land, land become sea. Where is the edge? Where does one end and the other begin? Grass, mud, and tide ebb, flow, and pulse together with the rhythm of my heart. With the subtle surge of tide, I too grow thick and luscious . . . in a final movement of contemplating the nature of self, I look to water, the tide and marsh before me. It moves, like me, spilling, shifting sand, the edge of land continually eroding. Pebble and stone, soil and sand, slip and tumble into new forms, new configurations. Grass and roots release, die, drift to sea, buried at sea, becoming one, the same. Sand, soil, marsh, grass, marsh, sea, the blood in me, I sweat salt. And pulsing like tide, I become the sea. For a moment, I become vast, expansive, deep. It is visceral. In the next moment I become gratitude itself, in quiet response to the possibility of such depth." Water, Sewall concludes, "speaks to us universally. It is a shrine, a calling, a catalyst for understanding the Tao. 'Learn to be water, let the moon strum your belly, the planets beckon and tug,' says the poet Morton Marcus."

Where does such a bonding leave the rest of us—the old, the young, the males? How else can we all go about reconnecting with the water world, together?

### Born Again

The reliance of environmentalists on doom and gloom as motivating forces for change has instilled a psychology of shame. In order to mature, environmentalism must shift its focus from reacting to, and therefore allowing itself to be defined by, the industrial-commercial complex. There is a need to develop proactive ways to embrace whatever it is that we most value in nature. There is growing recognition that the way out of the problem is to attempt to recapture memories of those feelings of aliveness, awakedness, well-being, excitement, exuberance, unbridled energy, joy, and above all else, the unfettered and non jaded hope through which we experienced childhood. We need to release the body's inherent playfulness and humor and to recover the child's innate ability to experience a world free of judgment. By celebrating life in this fashion, our collective depression may be ebbed. Nostalgia can therefore be a powerful means to invoke a return to water's joy.

Lynda Wheelwright Schmidt, for example, fondly recollects a childhood memory of when she and her siblings played like otters in a stream, followed by cleansing swims in the ocean: "We would head straight for the waves, naked as jaybirds . . . We would let the wash pull us back to the next wave and duck under it as it fell or dive through its curve, coming out behind it. A game was to line up in a row and ride a wave in as seals would, only our heads sticking out in the tube formed by the falling water, shouting to each other in the moment before we went under. The water was icy cold, however, and finally even we were too cold. Teeth chattering, we would race up to the hot, dry sand and burrow down into its delicious warmth."

Likewise, Howard Clinebell provides a letter from Tom Summers, a chaplain and pastoral psychotherapist who wrote to him "Many of my summer childhood and adolescent days were spent either in the depth of those cool, dark waters, exploring its rocky bottoms; or else dashing through the muddy swampland . . . I learned to swim in that river at the age of three, by first experiencing in its swift currents the opportunity to hold fast to my father's neck as he swam back and forth across the river, allowing me to experience trust and support in him and the water. Then there was the cold winter in my adolescence when four of my teenage friends and myself adventurously vowed to one another that, regardless of the river's cold waters, we bravely would jump into those freezing waters each month . . . My experiencing of the river has stayed with me very vividly as an earthy thread and is a deep part of my history. I now find in my adult living when I am in need of solitude or when

41

I might feel especially perplexed over deep questions, I often desire to be near water, like at the ocean, or some mountain lake, or some river. My adult needs in these respects are rooted deeply into the nourishment and the healing comfort found in the [water] . . . The thread with the river certainly gives me no answers, but its deep presence allows me, hopefully, to risk being more available and sustained as I experience the searchings and the questions of others in the human mystery of inner and outer voyaging in life."

In the end, there is a simple peace found in re-living our childhood through engaging water. "I ran forward like a child, tore off my clothes, and threw myself into the water" Stephen Swinburne recounts, before continuing: "It lasted for only a few minutes, but I was in heaven." Perhaps we should not be surprised at water's rejuvenating power. For as biochemists inform us, our bodily proportion of water declines from a high of about 78 percent at birth to only about 60 percent upon reaching adulthood. In other words, as H. G. Deming concludes, "we age because we become dehydrated."

### Hydrotherapy

Water has long been known for its imagined ability to renew the spirit and replenish the soul just as much as it rejuvenates the body and refreshes the appetite. The sensuous, restorative ritual of the Japanese bathing experience is an ancient tradition, for example. Water quenches us spiritually as well as physically: "O Lord . . . Have pity upon these waters within me, which are dying of thirst," says Paul Claudel. William Marks provides an extremely moving account of water's beneficial effects: "As I sat next to him in the wee hours of the morning, I would take a straw and submerge most of it into a glass of water. Then I would place my finger over the hole on the top end, lift the water-filled straw from the glass, and place the bottom end in my father's mouth. Slowly and purposefully, I would release the pressure of my index finger at the top of the straw. As the water trickled into my father's mouth, he would move his lips and tongue to help clear his throat. As amazing as this may sound, the last words he spoke before lapsing into a coma from which he did not revive—'Mmmm, that's good water.'"

Early Christian baptism involved immersion in a river, with "the power of the flowing current symbolizing the spiritual flow of life within the person's own body," as David Kidner describes it. The process was originally one of removing sins. Today's New Age dabblers can simply purchase a package of "Wash Away Your Sins" bubble bath and follow the tongue-in-cheek instructions: "(1) Kneel before thy tub; (2) Reflect upon wrongdoing; (3) Run warm bath; (4) Pour in enough bubble bath to equal your sins (double the amount you estimated); (5) Submerge thyself in blessed bubbles; (6) Soak; and (7) Arise cleansed from sin and ready to do it again."

The Harvard Library System lists 418 sources under the search for "hydrotherapy," spanning more than four hundred years, with most being concentrated in the early decades of the nineteenth century. Reading through the titles, a sampling of which is provided below, one is struck by the assured conviction of those promoting and defending the various forms of water treatment:

- *Water as Applied to Every Known Disease*
- *Principles of Hydrotherapy, or the Invalid's Guide to Health and Happiness*
- *Children, Their Hydropathic Management in Health and Disease*
- *Consumption, its Prevention and Cure by the Water Treatment*
- *Cholera, it's Causes, Prevention and Cure Showing the Inefficacy of Drug Treatment and the Superiority of the Water Cure in This Disease*
- *Outlines of a New Theory of Disease Applied to Hydrotherapy Showing That Water is the Only True Remedy*
- *Every Man His Own Doctor: The Cold Water—Tepid Water and Friction Cures as Applicable to Every Known Disease to Which the Human Frame is Subject*
- *Practice of the Water Cure with Authenticated Evidence of its Efficacy and Safety Containing a Detailed Account of the Various Processes Used in the Water Treatment*
- *Hints to the Sick, the Lame and the Lazy, or Passages in the Life of a Hydropathist*
- *Art of Swimming and Rules for All Kinds of Bathing in the Preservation of Health and Cure of Disease*
- *Water Cure for Ladies: A Popular Work*

- *Hydrotherapy in the Treatment of Disease by Water Applied to Warm Climates*
- *Sure Water Cure*
- *Discourses on Cold and Water Bathing with Remarks on the Effects of Drinking Cold Water in Warm Climates*
- *Medical Reports on the Effects of Water Cold and Warm as a Remedy in Fever and Other Diseases Whether Applied to the Surface of the Body or Used Internally*
- *Extract from an Account of Cases of Typhoid Fever in Which the Affusion of Cold Water Has Been Applied in the London House of Recovery*
- *Inquiry into the Efficacy of Warm Bathing in Palsies*

It would be wrong to suppose, however, that hydrotherapy is a quaint, all-but-forgotten relic from the explorative period of early health care. Today, the New Age movement has enthusiastically plunged into bizarre homeopathic "cures" thought to be brought about by water's mysterious healing energies or "memory." And *Watzu* bathing is a popular means of stress relief, allowing practitioners to flee the hectic bustle of modern civilization. Consequently, spas have, in the words of the director of the Monte Carlo Thermal Baths, reinvented the bath "as a source of solitude and renewal," where the body "is freed and surrenders itself, becoming more beautiful, more slender, more supple in the grace and equilibrium of this pristine element," the physical form being but "a mere envelope we cast off as we might a suit of armor made of stress and fear."

To complement her haunting photos, Linda Troeller collected testimonials, referred to as "spa stories," of modern believers extolling the manifold virtues of water immersion. For some, it is the sense of peace and security that remains the most vivid memory: "During Watzu lessons, I feel how the buoyancy of water holds and supports the body, while the mind becomes still. The isolation is like returning to the womb, carrying the body to a state of rapture." Others may be lost in a bliss of ecstasy wherein visions arise: "In a perfectly balanced saline pool I was bathing in underwater music and laser light. The songs of the Orca bubbled around me. I opened my eyes and found the other participants gathered like schools of fish, gliding on the surface, resting and sinking into deep blue, red, green hues of light. My body yearned to dance with the others. A yellow sun covered my face and a vision of whiteness followed." And for some, the experience can approach the spiritual: "A monk told me that the body is a source of transcendence. He had begun experimenting with water therapies. In changing the physical body he had radically changed his idea of spirituality. In cleansing himself, he was cleansing the Earth."

There is even a series of albums, *The Sound of Hydrotherapy*, designed for use with one's own bathtub at home. Album descriptions are couched in the flowery prose of New Age aspirations: *Still Waters*—"A harmonic sensation of peace and relaxation derived by the influence of music and rhythm leaving the mind and spirit to embark on a tranquil journey"; *Fountains of Youth*— "A revitalizing experience of musical suggestion that stimulates the fantasy of the listener, and regenerates the spirit of the mind"; and *Falls of Joy*—"An awakening of the psyche through dynamic sounds and rhythms, which recharge the spirit and soul." And if all that is not enough, there is even a book, *Water Magic: Bath Recipes for Both Body and Soul*, in which the initiated can design and blend their own healing elixirs into which to plunge.

It is, of course, possible to dissolve oneself into water in nature, far from either spas or home bathtubs. Alphonse de Lamartine found solace in water's cradling embrace. When lying in the bottom of a boat, enjoying "one of the most mysterious sensual pleasures in nature," he relishes the slow, smooth, and rhythmic movement of the gentle swells felt through the hull: "Water carries us. Water rocks us. Water puts us to sleep. Water gives us back our mother," he reflects wistfully as his soul is thus floated. At another time, while swimming, Lamartine imagines himself moving in pure ether "and being engulfed by the universal ocean." It is no surprise then, that the latest spiritual fix to hit the West Coast of Canada is "kayak therapy." Here, guests staying at a B & B called "Serenity by the Sea Retreat," go out with a "spiritual guide" to experience a New Age rebirthing ritual of paddling and healing.

This desire to return to some prenatal state of imagined comfort is a long-recognized attraction in water immersion. As Stephen Harringan writes: "Water is the human mind's most powerful mother symbol—a reminder of the benevolent, nurturing void from which we all emerged." He goes on to recount that pregnant women are known to dream of floods and ocean tides, seeing their children floating along the surface. "Even the letter M, the universal 'Mother letter,' is an ideogram meant to depict the movement of waves across the surface of water," Harringan concludes.

# Part II

# Wetted Appetites:
# Approaches

*It is from strenuous effort verging on the extreme limits that men and women seem to acquire a new power in themselves which in its turn leads to some sort of "total" view of things; as if exhaustion of their bodies freed their spirits from some clogging shackle of comfort and left them free.*
—Wilfred Noyce, *The Springs of Adventure*, 1958


*Exploration is the physical manifestation of the intellectual passion.*
—A. Cherry-Garrard, *The Worst Journey in the World*, 1922


*It is pleasant to hear the sound of waves and feel the surging of the boat,— an inspiring sound, as if you were bound on adventures.*
—Henry Thoreau *in* Robert L. France, ed., *Profitably Soaked: Thoreau's Engagement with Water*, 2003

# Adventure

### Thrill

As has been long recognized, part of the very thrill of partaking in adventure comes from engaging in a mildly threatening situation with abandon and satisfaction. Grey Owl aptly describes the anachronistic life of twentieth century voyagers as being "carefree and debonair, wild, reckless, and fancy-free . . . gallant, and rollicking" when they rocket down "the tossing foaming River." This is why for us today "paddling a canoe may be a kind of projecting ourselves into a simpler, more adventurous past" filled with many "romantic associations," as Edwin Teale contends.

The sense of danger lying just beneath the surface gives added weight to the import of such acts. Participants revel in their belief that somehow they have just skirted possible disaster. Some, like John Hildebrand refer to the "perverse thrill" of running rapids; others, such as Robert Collins and Roderick Nash, admit to having shaking hands following such episodes. Carolyn Servid, while lying in bed the night after a river dunking and reflecting on her bodily aches and pains, expresses gratitude that the physical discomforts were "enough to affirm a life that the earth had shaken and humbled and blessed." Swimming in treacherous rapids is for Roger Deakin an experience of heightened awareness, the underwater images magnified by his face mask, the sounds of the breaking water and his own breathing amplified at the same time. Likening himself to a soaring fulmar over dangerous seas, Chris Duff exclaims "Yes!" digging his kayak paddle into the sea and throwing up a bladeful of sparkling water into the air: "I was home, riding the power and lift of giant seas again. This is what I had come to experience."

It is that very instant—what William Least Heat Moon refers to as the "horrible pause" reminiscent of reaching the crest of a roller coaster or what Chris Duff likens to being offered up as a kind of sacrifice—that ?as inspired many. For it is here, with the first glimpse of the imminent descent into a possible watery oblivion, Moon's "yawning black hole" or Duff's "oily blackness of rolling seas," that one experiences intense, adrenaline surged feelings of a "stomach twisting sense of foreboding," in Chris Dreyna's words.

The downward rush can also be experienced quite differently. For Moon, it is with humor: "Wet your pants, cowboy?" his companion asks. Drenched with water, Moon replies: "How would I know?" For Dreyna, it is with a panicked cry: "Paddle!' I shouted above the din of crashing water." And for Jon Turk, riding inside the curl tube of a massive breaking wave having "no horizon, no universe, no future, just the water poised in an improbably graceful arc over my head," it is a time for oddly calm reflection on the likelihood of being dumped: "My mind stepped outside the situation and told me, 'Look around, Jon, remember this. You will never be here again.' It was an odd voice: quiet, unhurried."

### Danger

Much has been written about the psychoanalytical motivations behind recreational risk-taking. Often it comes down to the realization that, through living nuetured lives where the most dangerous undertaking for many is riding to work on freeways or subways, we have lost the sense of wonder, mystery, and especially awe of nature in its rawest and therefore truest form. In lives felt lacking structure, contrast, and occasionally even purpose, some feel the need to challenge fate by direct and deliberate exposure to situations fraught with life-threatening danger. Here danger represents much more than merely an easy means of thrill seeking; instead it becomes a vehicle through which lives themselves can be defined, understood, and even celebrated.

Once while scuba diving, Stephen Harrigan developed a fear familiar to all those who have ever donned a face mask and stared down into the dark depths. Letting go of a small starfish he had been holding, he watched it spiral downwards and then suddenly became nervous about maintaining his own neutral buoyancy: "The spell was broken, and I saw it was possible for me to fall through the water as helplessly as that sea star had. I had never been afraid of water, but now I stroked anxiously with my fins, realizing that there was nothing to support me in this element but my own exertions. I was like a bird beating its wings at the top of the sky. If I stopped moving forward, I would drop. Shore was fifty yards away, and I raced toward it, propelling myself not just with my fins but with my arms, my confidence growing every moment as I got closer to the beach and the sand bottom curved upward to meet me." Gaining the shore, Harrigan sat down in the rain and pensively wondered whether the unsettling experience had permanently ruined his joy of diving.

Recounting a sudden squall that shifted a situation from one where she was singing to one in which she felt "humbled and shaken," Ann Linnea learns the lesson of mindfulness: "Whether in ecstasy or storm, we must be totally present to the moment, ready for the sudden shift of wind or mood or energy." For it is through conscious fear of the loss of what is most significant, that we become aware of what is really most important. With return to safety often comes an accepting recognition of even those very mundane elements of our lives that drove us to engage in such extreme risks in the first place.

Interviews with survivors of near-death experiences reveal a surprising similarity in progression of emotions. First, there are feelings of fear. "There's no denying a sense that we don't belong here," Richard Nelson recounts as he is pulled toward a shore rocked by violent, crashing surf. Then come expressions of denial. "This is a dream. I'm not even supposed to be here," Tom Brokaw argues to an uncaring set of rapids. Next come feelings of "why me?" anger. "And later knowing I was almost certainly going to drown filled me with rage," John Hildebrand matter-of-factly states. Then, in desperation, with no other avenues of escape available, the doomed frequently begin to beg or bargain to be released from the cold, wet grip of fate. For some, like John Hildebrand, this leads to prayer: "Save me. Spare my life" he intones, and then, after surviving the rapids, continues: "I felt that sudden absolution of fear typical of takeoffs and landings, the quick prayers quickly forgotten." For many, awareness of how tenuous their situation has become finally registers with a sudden shock. "I'm gonna die," Tom Brokaw realizes; "I was going to die!" Dan Starkell exclaims. And Lorian Hemmingway, accepting her seemingly inevitable fate, rationalizes: "You are gonna die, I told myself. This is the exact moment you will die . . . So take it all in, baby. But things were streaking by too quickly to register. I would die remembering a blur."

Sometimes it is the objects harbored by water that bring the most fear. In a humorous exchange, Kevin Patterson debates with his sailing companion about the likelihood of encountering sharks before leaping off the boat, "testicles retracted up to my kidneys," rationalizing that "it's a pretty big ocean, what are the chances?" Scott Jenkins matter-of-factly states that for surfers "when you paddle out, you enter the food chain." Pursued by a potential gold-medal winning water snake, Lorian Hemmingway, herself admittedly only a bronze-medal swimmer, looks back once to see the head raised above the surface: "uncanny, the way he kept a bead on me, adjusting his movement with each stroke I made, the dead look in his eyes the exact dead look sharks have when they're feeding." And while paddling in the South Pacific, Paul Theroux feels a sudden thump against his kayak and wonders if he has meet a shark attracted by the thrashing paddles: *"Here is a panicky creature that is afraid of me*, the shark reasoned instinctively, *I think I will eat it."*

## Hardship

Water can be experienced in both a benign relationship as an idyllic and rejuvenating baptism of the soul, and also in a relationship that tries the fortitude of the human participant. In the latter, the struggle against water is used as a means to explore strengths and limitations of human nature which otherwise would have gone unrecognized. Similar to the rationale that many profess for running marathons, this is the "that which does not kill us makes us stronger" school of growth. This approach is based on a personalized view of water as a testing ground toward seeking self identity, similar in spirit to Mallory's famous bathos about climbing Everest—"because it's there." By forcing us to the very limits of our abilities, both physical and mental, water plays a utilitarian role in fostering human fulfillment. Such an outlook needn't invoke water as an antagonist, and indeed most writers do not. Again, to use the analogy of mountain climbing, Mallory's statement of intent is appropriate with respect to water as well: "Have we vanquished an enemy? None but ourselves."

The literature on water is filled with accounts that read like litanies of masochistic torture rituals. For example, R. M. Patterson refers to the "half-drowned sputter" of his companion desperately clutching to a safety line with arms "pretty nearly wrenched out of their sockets." David Quammen vividly remembers time standing still as he too hung in the water "like a dipstick," holding his breath until almost blacking out. And when a rogue wave hits Chris Duff, it leaves him shaking "like a wet rat" as it tears at the paddle with "the fury of a watery tornado," filling his ears and sinuses with stinging cold seawater. William Least Heat Moon struggles with arms aching from holding the steering wheel until he and his companion "stand helpless and stupid like cattle and just get beat" by a raging storm that leaves them with blue lips and chattering teeth, exclaiming "hypothermia here we come." And Ann Zwinger describes the toil taken on one's legs during the process of lining a canoe down a rapids: "ankles crack and shins bark against unyielding stone . . . it is simply a head-long progression into disaster with every foot for itself."

Jon Turk's account of struggling to make a beachhead in Patagonia is particularly memorable: "My feet touched the sand, the undertow tugged against my body, and I lost footing. A big wave approached, and I hung on to the kayak. The wave lifted the boat over my head and effortlessly yanked my right shoulder out of its socket . . . Certainly my life didn't pass before my eyes or anything romantic like that. I mentally crawled inside my body and tried to picture the positions of the bones and ligaments . . . With my left arm, I grasped my right elbow, rotated the shoulder, lined up all the parts and pieces, and shoved it back into its socket. The pain was intense. I rolled out from under the boat and staggered onto the beach."

Grimace as we might when reading such accounts, this sub-genre of water writing reaches its apogee in describing the feelings of complete helplessness that often accompany adventurers who literally and figuratively find themselves "in way over their heads." By portraying their futile efforts in extracting themselves from the power of the threatening liquid embrace, while at the same time describing the toil exacted upon their bodies, these writers vividly convey the sense of being sacrificial offerings to the whims of water's destiny.

Carolyn Servid recounts how after being immersed in a strong current, her body is slammed repeatedly against the boulders "in the murky moving cold" while she struggles "uselessly" to right herself before she is swept "uncontrollably" downstream: "The force of the water carried me along a sacred edge where life meets its source and its beginnings. . . I remember struggling and struggling and then a moment when, exhausted beyond my means, I felt as if I could just slip into sleep." Lucky for Servid, she finds refuge on the "holy piece of ground" of a sandbar where she collapses and weeps. Others, however, are not so lucky to escape with just bruises and a scare, as this account of how Dan Starkell lost most of his toes illustrates: "I surged forward with no hope of lasting. My shins and knees smashed against boulders below the surface, deadening my already stiff and cramped legs. Scared, numb and useless, I clung on to the kayak being shoved slowly towards those boulders that I would never reach. All of a sudden, the lights went out and I was gone… Four more times I broke through the honeycombed ice, up to my armpits in icy water many fathoms deep. The dread of being stranded, with the shock of ice-water on an already chilled and terrified body, was overpowering me. Hypothermia was setting in fast."

### Adversary

There exists a long tradition in nature writing wherein protagonists are portrayed as waging a heroic battle against the harsh forces of an uncaring, and even occasionally malevolent, nature. Here, it is man (rarely "woman") pitched in a war in which human ingenuity and strength are put to the test as never before. Nature does its best to wreck ruin, but with fortitude and good luck, the wise and well-thewed human wins the day, surviving to triumph his victory over nature in retributive prose. Although such testosterone-filled encounters with nature may seem simplistic and retrogressive, they nonetheless still represent a major influence on how individuals intimately experience water.

The adversarial interaction between man and water is summarized perfectly by Grey Owl describing one of his many canoe explorations of boreal rivers: "We must fight the current, to escape it and catch an eddy, for just ahead is a standing rock against which the full force of the River hurls itself in unforgivable fury, striking with terrific impact . . . for this is Men against the River."

More recent writers are no less likely to describe their encounters in similar combative prose. William Least Heat Moon refers to "struggling against a river that rewrites mile after mile every law of hydraulics yet advanced by science," while having to wrest control of a recalcitrant steering wheel from the water's pull. For Robert Collins and Roderick Nash, a dangerous river run is likened to a billiard game between the rubber of their boat and the hard unyielding surfaces of the rocks, whereas Tom Brokaw simply refers to his white-water experience as being "a battle for survival." Chris Duff survives a series of large waves along the Irish coast only when the seas "relinquished their hold," allowing him to retreat into the welcome calm of a river mouth. And Dan Starkell's kayak paddle through the Northwest Passage reads as one long championing of machismo as he struggles to wrest another few kilometers' passage from the ice floes that the Arctic seems to go out of her way, he believes, to strategically position right across his path.

This latter strategy of personalizing nature as a wily antagonist doing its best to circumvent the noble plans of the puny human hero is a well-used dramatic tool. For Craig Childs, a waterfall pounds his head "like fists" or like fumbling "mad hands" trying to pull him down. Rarely has this sub-genre been invoked with such weighty passion as by David Quammen in his likening of the water's force to "a pissed-off Old Testament God, say, or a six-foot-eight mugger." At one stage, having nearly had his paddle wrenched from his hands, he hears a chiding voice from the turbulent rapids: "*All right, chump, I want your paddle and your glasses, yeah, and your helmet and your life jacket as a matter of fact, and I want that gold off the back of your front teeth. I want your boat. I want your booties and your shorts and I want all your money. After that, we'll consider your life.*"

*Happiness, I think, is a simple everyday miracle, like water.*
— Nikos Kanzantzakis, *Japan, China*, 1963

*That the water was wont to go warbling so softly and well./ How good is man's life, the mere living! How fit to employ/ All the heart and the soul and the senses for ever in joy.*
— Robert Browning *in* Wilfred Noyce, *The Springs of Adventure*, 1958

*Immortal water, alive even in the superficies, restlessly heaving now and tossing me and my boat, and sparkling with life!*
— Henry Thoreau *in* Robert L. France, ed., *Profitably Soaked: Thoreau's Engagement with Water*, 2003

# Joy

### *Excitement*

Modern lives, free from the many daily hardships of our ancestors, have become commonplace, predictable, and often exceedingly dull. We collectively suffer from the "C-syndrome": complacency, comfortability, and contentment. Vaguely remembering some faint thirst arising from deep within, we go out of our way to attempt to spice up our mundane existences. For some, those whose inner spirits have been deepest buried, fulfillment of this need never proceeds much further than that conveniently offered up on our television or movie screens or at sporting events. For others—the more adventurous, thrill seeking and the purposeful exposure to the unknown is a hugely popular pastime. It is no accident that as our daily lives have become increasingly stale, people have increasingly turned to the outdoors for their adrenaline fixes. Moving water, even more than soaring summits, has become the major recreation vehicle for experiencing the excitement that nature can offer.

There is a long-established tradition in seeking the thrill of running rapids. The stimulus of fear, the flush of confidence, and the rewards of independence are all important motivating forces. In a passage that ends with many exclamation marks, Grey Owl captures this spirit: "So, speed, speed, speed, grip the canoe ribs with our knees, drive those paddles deep, throw your weight on them . . ." Sigurd Olsen is drawn to rapids for the feelings of "excitement, eagerness, and joy" that they instill. With each set of rapids successfully navigated, Eddy Harris gains "more and more confidence," leading him to "shout with triumph and glee." Similarly, after one particularly tricky set of rapids, Jeff Wallach enjoys the pride of the moment: "I scream as loud as I can—a whoop not so much of victory as release, draining the tension that has built up in me, releasing it into the cool, wet air." Ann Zwinger, though being "apprehensive" and "trembling" at the start of a river trip, later writes: "The rapids begin with the rushing sound of wind through wings. Then the bow lifts and we soar with the river," and "The boat swings and pivots. I hold my breath. I feel as if the top three layers of skin are gone in a total exhilaration of water streaming down my face, down my neck, totally soaked in spite of a rain suit, disembodied yet physically involved, aware of every rapid I've ever run."

River runners often regard their boats—the means through which they communicate with the water—with great pride and near reverence. Referring to the "rambunctious grace and style" of river dories, Jeff Wallach lovingly describes how such a craft "seems to breathe beneath you as you talk it through rapids," and how "these boats come alive in whitewater, as if they're actually ecstatic to be there. They practically sing." Ann Zwinger's boat also seems alive: "The canoe tugs beneath my hand. It seems animate, eager to move, listening just as I do to the sound of more rapids jubilating downstream."

The exhilaration of experiencing water is by no means restricted to rivers. Richard Nelson, exploring a sea cave with his kayak, navigates the choppy rebounding waves with enthusiasm: "My whole body tingles with adrenalized fear and excitement, as the shining mound of water comes inexorably toward me." And Jon Turk exclaims, "Yahoo, technical sea kayaking!" after being spun around in eddy currents. In one humorous escapade involving slipping and sliding over polar ice as they pull their kayaks behind them in attempt to catch the retreating tide,

Don Starkell remarks of one of his companions that "it was the first time he had found himself running for his life, dragging a kayak in inches of water while five miles out to sea, trying to eat lunch and hold up his pant legs, while at the same time trying to have a pee."

And of course one can appreciate the excitement of water perhaps best by actually being immersed in it. Edward Abbey enthusiastically describes the "cheap thrills" of swimming in rapids: "The waves soar above your head, blotting out the sun. You gasp for air just before the water wallops you in the face, rise into the light at the crest of the wave, and descend like a duck into the next trough." In the same vein, J. Calvin Giddings writes: "Suddenly I burst out the other side like a champagne cork, and felt slick rock at my foot. I groped for a place to stand, reached up to the edge of the formation, and pulled my dripping frame out on a gently sloping platform of pink rock. 'Yah-hoo!' I shouted in ecstasy, and my voice echoed and reechoed 700 feet up the shadowed walls of a desert gorge."

### Relaxation

In contrast to the thrills-and-spills school of many modern outdoor recreationists, others remain drawn to water simply to enjoy the delights of slow moving cruises. There, with the pressures of office and family lives temporarily left behind, participants revel in the immediacy of the moment, savoring feelings of peace while being carried with carefree trust upon the bosom of the placid water. Ann Zwinger and Edwin Way Teale, for example, use a trip into the inner sanctum of a wetland as a simple means of escape from the hectic world: "Slowly the surf sound of the highways receded, sank into a seashell murmur, then ceased completely as we wound deeper and deeper into the swamp." Likewise, Chris Duff exalts in the ease of moving atop a perfectly still sea, his paddle blades "slicing into a mirrored sky" with each stroke as he lets comfort soak into him, blending with the undramatic but "healthy strain of muscles against waves." In many respects, engaging in such a pursuit is a much more mature activity than that sought by the thrill-seekers dashing from favored location to location, never pausing long enough to enjoy what the slow stretches of water in between can offer. In another passage, Duff specifically states that it was not adrenaline he was after, but rather a desire to be part of a wider world, as he savors the joy of a memorable day when "the forces of the sea and the winds combine and flow together in near-prefect unison."

Edward Abbey captures the sublime placidity of lazy drifting so perfectly that one can easily feel the languid moment: "The cool water flows between my fingers. My kid-daughter plies her oars three boat lengths ahead, serenely delighted by everything. My friends lie sprawled on their boats beyond, floating and sunning and dreaming." Kathleen Dean Moore describes a blissful evening paddle under a sky filled with shooting stars as a kind of metaphor for the hurried lives left on the shore, savoring the peace and safety of the water refuge: "We checked the sky to see if any stars could be left. We listened for trumpets. All this glorious disaster and not a sound but the canoe pushing softly through the warm water and Frank, quietly reminding me to paddle."

For Kevin Patterson, the ocean is an escape from the otherwise enveloping trauma of heartbreak. When the boat's engine is finally turned off the first night at sea, he realizes with a shock that in the intense darkness and "glorious quiet," they are in the presence of something larger than anything he has experienced before: "We rose up with the swell and then down, and it was like a hand was underneath me, holding me just a few feet above the water." Carolyn Servid's boating sojourns are likewise motivated by a desire to experience the ocean in just such a way: "Sometimes I rowed out just to sit on the water and feel the constant pulse of its undulations." Thus, one can achieve solace, if not true inner peace, through a perspective of losing of one's smaller troubles in the vast expansiveness hinted at in the regular rhythm of waves.

### Playfulness

Water, more than any other element (except, perhaps, water as snow), instills a feeling a playful glee, allowing participants in the aquatic dance to be temporally transported back to a time in their childhoods when splashing about in puddles on the walk home from school was an unquestioned, completely natural act. As children, of course, we could never understand what crime had been committed to merit the parental displeasure when we arrived at the doorstep soaking wet from such puddle peregrinations. Now, as scolding adults, most of us have forgotten the powerful magnetism through which those puddles exerted their hydrophilic pull. Other adults, however, have been successful in using the time machine of water to relive their youth.

Edward Abbey recounts a wonderful water fight among warring canoes drifting down a languid river. Likewise, John Jerome writes, "I was eleven years old again, the river as delightful and the world as wild, endearing and rich with possibilities as it had ever been." For Janet Lembke, skinny-dipping is a "time to get exhilarated" when she and her companion shed their "clothes and grown-up lives." "Children again, giggling with residual naughtiness but not caring if anyone sees or hears, we launch our bodies forward, blending sweat and brine," she writes. The illicit pleasure of playing with water, this time while sailing, is also experienced by Kevin Patterson: "I felt as if I were riding a bike no hands with my mother watching. Some juvenile gesture of bravado. It felt preposterous and absurd. Extravagantly ill-conceived."

Childhood is a time when less important realities are squeezed between serious games of fancy. Water offers a return to such a period. John Malo, after a day of skating on a river, plays shinny into the wee hours, and builds floating walkways out of ice blocks. Sigurd Olson, actively canoeing into his old age, nostalgically reflects back on the "sheer fun" of his boyhood river encounters, those "joyous adventures" which gave him and his companions "plenty to laugh about." Craig Childs and his companion plunge into a desert pool and "instantly we become playful, embarrassingly so as we made the water slosh back and forth unnecessarily until it tipped out one end." And like a child sloshing down an amusement park waterslide, Roger Deakin repeatedly plunges into a tidal bore that carries him speedily along in what he refers to as "dream swimming."

For some, challenging encounters with water are often approached as if engaged in a game. Ann Zwinger recounts how after losing a tug-of-war and being pulled down a steep, muddy bank into the waist deep water while clutching a tow rope, she started laughing so hard that she couldn't get up. Pulling kayaks over a series of ice-floe obstacles becomes a "jigsaw puzzle" for Jon Turk and his companion: "Waves soaked our pants but the wetness was irrelevant; everything was irrelevant except this childish confrontation with one piece of ice amid a continent of ice. When the floe finally broke, I lost my footing and fell into six inches of icy water. Chris laughed."

Encounters with aquatic biota while on or in water have also been approached as a celebration of the joy of life. Richard Nelson symbolically participates in the sweeping flights of birds by gamely playing in the waves beneath them. And Jon Turk attempts to become one with a group of dolphins: "The game enforced my paddling rhythm; if I missed a stroke or took one too late, I would hit a dolphin on the back and that would be a rude thing to do to one's playmates. I also realized that if the dolphins lost their rhythm, one might sever my flimsy kayak with its sharp fin. But they were born in the water, had evolved aquatic precision for millions of years, and understood the rules of the game."

## Happiness

Water, the element of life, sustains not just through direct ingestion, but also by nurturing the inner spirit. In other words, to deeply experience water is to live to the fullest. Grey Owl recognized this when he wrote, "We all laugh and join in the chorus. We begin to enjoy ourselves, to rejoice in the fluid rhythm of the canoes, to feel the ecstasy of this wild, free, vigorous life that seems all at once to be the only life worth living." Once, when observing a pair of singing canoeists, Sigurd Olson reflects upon such a tradition: "I knew within them the long inheritance of a nomadic ancestry was surging through their minds and bodies, bringing back the joy of movement and travel, adrenaline pouring into their veins, giving courage to muscles being strained to the utmost. If I had been close enough, I might have heard the laughter in their song, seen the glad light in their eyes." Ann Linnea explains how easy it is to be drawn into song through the tempo of paddling: "My voice was timid, quiet. But the more I sang, the better I felt. As my confidence grew, my volume increased, my paddling became more rhythmic, like a dance. I was dancing with the waves, singing to their tune. Paddling like heaven." Chris Duff feels pressure to release his happiness into song: "Again I sang. I had to. I had to lift my voice, mix it with the winds and the seas, and offer it to the wildness and the pageantry that surrounded me. It was either sing or explode like the waves, vaporized by the energy that my body could not contain. Each time a massive roller collided with a reverberating whumph against a hidden rock or sea stack, I would sing out, 'Yes! More, more. Yes, you are beautiful!' The seas filled me, they rolled through me, and built into breakers of emotion that erupted into shouts of ecstasy. I was caught up in the passion, a wild man singing his heart and voice to the sea and the winds." After a refreshing dip into a desert pool, Craig Childs's companion, water dripping from his body, falls back to the hot rock: "He closed his eyes. His mouth opened. In a groan, he said Yes."

Water instills feelings of aliveness of such intensity that individuals are forced to find voice in exuberance: "Whee!' I holler back. My shirt-front is soaked this time with deliciously cool, clean water, not salty sweat," Janet Lembke exclaims about a rain shower before continuing: "It's as if we walking, flying, wriggling, hopping creatures of the land have needed water-laden air as much as sea creatures need oxygen-laden water. Now we all celebrate the quenching of our heart's thirst. Drops pelt on the world like grains of rice flung at a bride and groom." "YAHOO!" Ann Linnea shouts after riding through a series of rolling waves. While kayaking, an "ecstatic" Richard Nelson hoots "breathless-ly," thanking the ocean for its "gift" of sea spray: "I am utterly, perfect-ly, exquisitely in its path, struck by a blinding maelstrom of salty mist, blown backwards and nearly capsized, drenched, breathless, and ecstat-ic. I've never been kissed so vehemently. It's moments like this that fire the deepest passion for being alive." In a similar tone, John Jerome writes that while swimming, "some unnameable voice whispered in my head, as I slid along under the surface, *Alive! Alive!*"

Perhaps, above all else, it is water's role in fostering fun that is most memorable. There is no need to intellectualize such an experience, as John Jerome recognizes: "Thrashing our way down the occasional rif-fles, capsizing more than once—with great hilarity—we learned almost nothing about how a river works. Whitewater enthusiasts talk about reading the river, but we hardly noticed there was a text. All we were doing was floating down our local stream." The sense of pleasure engendered by exposure to water, whether based on a search for excite-ment, relaxation, or childish playfulness, often produces a joyful bliss. For Ann Zwinger, "the lilt of the happy water" mirrors the sheer exu-berance and happiness she feels within for simply being on the river. Carolyn Servid explains her lingering on the water during evening pad-dles as a result of an inescapable urge to lift her voice in song. And when sailing through a choppy sea, Jonathan Raban, flush with adrenaline, exclaims in sudden and simple realization, that "*I like it here.*"

*The call of water demands, as it were, a total offering, an inner offering. Water needs an inhabitant.*

> —Gaston Bachelard, *Water and Dreams: An Essay on the Imagination of Water*, 1999

*We are surrounded by so much steel, rubber and electric wire that we should not sneeze at the chance of making direct contact with a natural element essential to life—the sea.*

> —Jacques Cousteau, *The Silent World*, 1953

*Now I forgot that I had been wetted, and wanted to embrace and mingle myself with the water.*

> —Henry Thoreau *in* Robert L. France, ed., *Profitably Soaked: Thoreau's Engagement with Water*, 2003

# Contact

### *Feel*

Like so many doubting Thomases, for some of us there is a physical need to grasp, poke, twist, and touch nature in order to fully acknowledge and appreciate its multitudinous splendor. Nowhere is this more easily accomplished than with water. For it is the receiving essence of the element itself that provides a satisfying sense of reciprocity while engaging in the tactile act. Other elements merely tolerate us; water intimately receives us.

The simple act of extending a hand toward water can be an act of honored penance, reverent supplication, or cautious exploration, in all cases the fingers acting like siphons to draw water up into the body. Michael Delp savors the numbing pain of offering his warm hand to the frigid embrace of a winter river. Bill Green recalls a friend being drawn like a sleepwalker to the toe of a glacier where, in silence, he stands on tiptoes, spread-eagled with outreached arms pressed against the blue surface of the ice. "I dabbled at the surface of this tiny pool with a finger as I had done to each one," Craig Childs recounts, "not out of conscious choice but involuntarily to keep from weeping, to do anything so that I could touch this water."

For some, it is as if their hands were plunged into the very life marrow of the planet. This is what motivates John Jerome to bury his hands deep into the sand and feel the swirling pressure of water emerging from a spring, or for Brenda Peterson to touch the sea bottom and feel the pulses of the waves pass through her in a strange familiarity that overwhelms to the point of rapture, or for Craig Childs as he endeavors to reach deeper and deeper into a desert spring, motivated by the desire for the "clandestine enchantment of encountering something small and sacred." And it is oars for rower, Craig Lambert, that connect him with the primordial waves arising from the depths.

Often, though, the simple insertion of a limb into water is not enough, there being no substitute for total bodily immersion. Roger Deakin reflects that cold mountain spring water "sends our blood surging and crams every capillary with a belt of adrenaline, dispatching endorphins to seep into the seats of pleasure in body and brain, so that our soul goes soaring, and never quite settles all day." John Jerome describes how once immersed in the "hugging machine" of water, our bodies become "all nerve endings" with the consequent and "ultimate firing of the pressure receptors." For Jerome, the pressure that water enacts upon the entire body—"it's gentle resistance to movement"—has the added benefit of sensorily locating one in space in a way denied in the terrestrial world with only our feet to anchor us. The irony, of course, is that in order to become truly grounded, one must become immersed.

To move from just touching water to really feeling it, is to move from mere pleasure to complete ecstasy. For Janet Lembke, water "sparkles, and swirls, surges and lapses. It splashes, rushes over rocks, whispers, breaks, and bubbles. It fills noses and mouths with a bromine reek, a coolness of trout and algae-covered stones, a tidal potpourri of seaweed, fish, and salt . . . it drenches our bodies, every pore, every nerve, every last cell, with the purest pleasure." To skinny-dip, she later says, "is to drown in bliss." After a cold swim, Roger Deakin explains that "if you tread on air on the way from the pool, it is because you are floating somewhere just above your corporeal self." Richard Nelson informs of his "great revelation" in learning as a child that clouds could touch the earth in the form of fog. After this epiphany, he would wander outside and enter such fog in order "to be in it, to feel its wetness, as if I'd magically walked up into the sky."

Roger Deakin, an individual obsessed with swimming as few others have been, philosophizes that through immersion one crosses boundaries and achieves a kind of metamorphosis with survival displacing both ambition and desire as the dominant aim. Such an activity, he argues, is the most intense and complete way we can be truly in nature.

### *Rejuvenation*

The solace provided through contact with water has been recognized for ages, as has the cleansing ability of water, washing away the accumulated worldly grime from both our outer and inner surfaces. It comes as no real surprise, then, that contemporary nature writers share a striking similarity in their descriptions of such baptisms. A rainstorm for Carolyn Servid provides renewal, the rain being "earth's sacrament, consecrating all living things." John Jerome recounts his lucky discovery early in childhood of managing to escape from the stress of the dry-land world into "a calm, quiet ocean of peace" that today, as an adult facing more trying stresses, is "still the ultimate tranquilizer." Swimming upstream, his mouth opened wide to gather in "as much water as possible," provides immediate succor to Michael Delp: "Wounds I thought I had forgotten suddenly heal, something inside my life gathers itself, turns further inward, lets the river pass through." And for Ann Zwinger, the simple act of standing ankle deep in bone-chilling water "without civilization, without defense, going back to self" is enough for her to be submerged "psychologically to the base of being," and in so doing, achieve peace. Roger Deakin feels revitalized after a swim in a cold stream due to its embrace resembling "mothers soothing and kissing us cool." For Craig Childs and his desert hiking companion, standing at the base of a waterfall is an electrifying experience: "Our arms reached upward, into the rainstorm, brushing the luxurious undersides of maidenhair ferns, pulling rivulets down to our faces and chests and legs, turning our bodies into connection points, like spark plugs or lightning rods. We carried water."

Deep immersion for some—to the very base of their being as Ann Zwinger would say—can be an act of restoration and rebirth. To "restore" implies a return to a time or state of greater purity. Mircea Eliade believed that because water in a sense dissolves time, it "signifies regression to the preformal, reincorporation into the undifferentiated mode of pre-existence." Immersion for Stephen Harrigan, for example, is a chance to "recede into the sleepy, safe, nonthinking being I used to be, before I was expelled from the womb and supplanted by my conscious self." In Terry Tempest Williams's case, this process provides a means to reconnect with the world: "We can always return to our place of origin. Water. Water music. We are baptized by immersion, nothing less can replenish or restore our capacity to love. It is endless if we believe in water." "To swim is to experience how it was before you were born. Once in the water, you are immersed in an intensely private world as you were in the womb," rhapsodizes Roger Deakin. Such an impression is reconfirmed when later as he slides down a series of waterfalls, Deakin states, "The slippery blue-green wetness and smoothness of everything, and my near-nakedness, only made me more helpless, more like a baby. It was like a dream of being born."

The "magical moment" when his eyes opened underwater for the first time shook Jacques Cousteau to his core. "I have been unable to see, think or live as I had done before," he wrote. As he floated weightlessly through space, "the water took possession" of his skin, and with it, came the conception that the effortless movements of the marine life about him had "acquired moral significance." In an epiphany, Cousteau realized that gravity was the true original sin, "committed by the first living beings who left the sea." Redemption, he believed, could only come by a return to the sea as already accomplished by the whales, dolphins, and seals. It is this same spirit that causes an already waterlogged Roger Deakin to hurl himself into yet another dodgy English waterbody, this time to crawl his way through the wet mud "feeling deeply primeval, like some fast-track missing link in our evolution for the lugworm," as he re-enacts what he refers to as "the evolution of swimming." Totally enjoying the situation, he wonders if he might have discovered some "new form of therapy; something along the lines of the primal scream."

*Eros*

Water is undeniably sexy—think Anita Ekberg in the Trevi Fountain in Fellini's *La Dolce Vita*. Sensually, it simply feels good to be cradled and cuddled all over by the searching presence of water. Sensuously, given that by its very nature sex is a wet act, how can immersion in a liquid embrace not be looked upon as being anything other than a little naughty? Because of this, water has long been a staple of the pornography industry, as, for example, an illustrated quarterly called "Wet Dreams: The Magazine of Erotic Water Games". With the recent publication of a collection of short stories advertised as being "the first waterproof book for adults," water has come out of the steamy back rooms and sticky magazines and into the front pages of serious erotic literature. For most though, it is not so much the act of making love to someone else in water that has resonated in their lives, as it is the somewhat odd act of making love while alone *to* water that has been most important. "My lips will feast on the foam of thy lips," the poet Stephen Swinburne wrote to a curling wave many years ago. Paul Valery stated "to twist about in its [water's] pure depth, this is for me a delight only comparable to love." And Flaubert referred to such carnal acts as "fornication avec l'onde."

For some, water eroticism is subtle, the imagination left to guess about intent and effect. Ann Zwinger simply refers to the "hedonistic enticement" of rivers; Thomas Faber, to the "erotic pleasure of surfing ocean swells." John Jerome likens drinking from a forest spring to "kissing" its surface "as wood sprites are supposed to do." Michael Delp describes the uncontrollable "urge" to lie upon the bottom of a river "and make love" by curling fingers into the gravel, taking a deep breath, and "giving yourself over." Stephen Harrigan confesses that the "cratering, yearning emptiness" he feels when staring at the blue tropical ocean is really a "sexual longing." Ann Linnea describes the "exquisite" feel of water and how her body responds with tenderness, sensuousness, arousal and finally ecstasy. In one episode while riding a series of large swells, Linnea admits to becoming intimately "seduced" by the "Lake." "Over and over I was massaged deeper and deeper into the mystery of connection with this gigantic being," she writes. John Jerome touches upon the reciprocity important in any relationship: "Clearly, we who love water feel we are loved by it: dive into it and it touches you all over, instantly, in your most private places. The ultimate mouth, or womb. It seems to love your skin; your skin certainly loves it."

Other passages about water's erotic effects are more explicit. Michael Delp anthropomorphizes water, in one case describing how when extending his arms into a deep, clinging fog he imagines caressing the breasts of a mysterious female whom is Water embodied. And in another very beautiful passage, Delp likens the river flowing by his cabin to an aqueous lover: "Under the sky, under the bed, under the house, the most beautiful woman I have ever seen is stepping out of her skin, as if out of delicate silk. She holds her skin in her hands as if it were cloth and begins to wring it slowly, and slowly, the most beautiful water begins flowing. When she lifts this water up into this world, her hands cup toward my face and when I drink her, I know for the first time that her river is where I have lived my whole life."

The eros of water is of course felt most intimately when participants are submerged. John Jerome comments on the "certain titillation" offered by water pressing itself upon one's various body parts, marveling that the act is "not only legal but is even socially acceptable." Ann Linnea "surrenders" herself naked to her anthropomorphized lake-lover. After interviewing many about swimming naked, Janet Lembke sums up by stating that "almost everyone agrees that skinny dipping is the most completely sensuous experience available to humankind." Roger Deakin enjoys the subversive activity of unsupervised swimming, lingering in the water to the "knife-edge between aching and glowing."

John Jerome informs us that female marathon swimmers are well known for indulging in "steamy fantasies during long swims," but believes that similar fantasies for others always stop at eroticism before becoming orgasmic. Not so for Terry Tempest Williams, who after stating that "desire begins in wetness," confesses to longing to be wet herself. In a passage of incredible passion, Williams shares an account of water masturbation: "I dissolve. I am water. Only my face is exposed like an apparition over ripples. Playing with water. Do I dare? My legs open. The rushing water turns my body and touches me with a fast finger that does not tire. I receive without apology. Time. Nothing to rush, only to feel. I feel time in me. It is endless pleasure in the current. No control. No thought. Simply, here. My left hand reaches for the frog dangling from my neck, floating above my belly and I hold it between my breasts like a withered heart, beating inside me, inside the river. We are moving downstream. Water. Water music. Blue notes, white notes, my body mixes with the body of water like jazz, the currents like jazz. I too am free to improvise."

## Taste

For many, it is not enough to merely immerse one's self into water; these individuals must in turn reciprocate the act by imbibing the water into themselves as well. There, working its imagined magic deep within the inner recesses of one's being, it may be possible to experience bliss, serenity, and a reprieve from a primal thirst that has hitherto gone unquenched. This is the most intimate form of baptism, one described variably by initiates.

Some have attempted to explain the sensations of water ingestion. John Jerome insists that if one dives into a lake, mouth agape, the feeling of fizzy carbonization is such "that your naked body will swear that your mouth is not telling the truth." Bill Green, a limnologist—a scientist who studies freshwater—"greedily" gulps water from a frozen Antarctic lake, "savoring its taste which was the absence of taste," allowing the cold to coarse inside of him, "clear as a crystal going down." For Michael Delp, drinking from a river allows one "to come as close as you ever have to purity," the river water silently intermingling with your internal bodily fluids toward "an invisible center" where you can actually hear "the sound of your own blood pulsing close to the river."

That Green uses the adjective "greedily," implies that an act bringing such joy might also be regarded as being somewhat untoward. Like other acts of pleasureful self-stimulation, drinking fresh water can be approached guiltily and surreptitiously. Janet Lembke, for example, recounts catching one such offender: "On the next walk, Sal remembers where she found refreshing drink, but she doesn't show me. The time after that, though, I catch her in the act. The place is right at trailside, a short hop and jump past station two. From shoulders to rump, she is visible, but her head is thrust between two slender tree trunks. And the sound of her lapping is loud and fluent."

The choice of words Lembke uses—"lapping" and "rump"—is designed to instill an animal quality to the scene. It is possible, she seems to be insinuating, that the act, the desire, to consume natural water is no different in humans than in it is in other denizens of the forest. Perhaps it really is a basic instinct in us all that occasionally manages to break free from our adult fears of contamination. John Hildebrand describes a beautiful encounter that certainly seems to indicate that this supposition might indeed be true: "The baby meanwhile had edged closer to the river. He stared at the water lapping against the shore as if encountering for the first time a new element, one that moved in smooth sheets of reflected light. His mother got up from her chair to bring him back into the circle. But before she could reach him, he lowered his blond head into the river and took a long drink."

*To contemplate water is to slip away, dissolve, die . . .*
    —Gaston Bachelard, *Water and Dreams: An Essay on the Imagination of Matter*, 1999


*In his first journal Thoreau entered this quotation from a French diction-ary: "From the primitive word Ver, signifying water . . . is derived the word veritae; for as water, by reason of its transparency and limpidness, is the mirror of bodies—of physical etres, so also is truth equally the mirror of ideas—intellectual etres, representing them in a manner as faithful and clear, as the water does a physical body."*
    —Robert L. France, ed., *Reflecting Heaven: Thoreau on Water*, 2001


*If one would reflect, let him embark on some placid stream, and float with the current.*
    —Henry Thoreau *in* Robert L. France, ed., *Profitably Soaked: Thoreau's Engagement with Water*, 2003

# Contemplation

### *Dissolution*

The feelings engendered by the best environmental writing are those of personal identification with the naturalist as s/he describes coming to recognition of being part of a grand union transcending the narrow and impoverished universe of the individual. The opportunity to forget, if even for only a brief moment, the division between the world of the outside and that of the inside, is a key insight into the very fabric of existence; and one that metaphysicians have investigated for centuries. Water, more than any other element, allows this to happen. "Swimming, like opium," writes Charles Sprawson, "can cause a sense of detachment." Indeed, this is what is clinically referred to by psychologists as being the "oceanic feeling." The idea of merging or blending with nature, losing the identity of one's selfhood, can be accomplished by deep immersion in water, the original lifeblood of creation. Many have attempted to dissolve themselves completely into water, and by so doing, paradoxically, have found their own true selves.

The search for achieving oneness with water has had many voices. On an ocean sailing trip, Martin Ames describes how he and his companion ceased to be alien "selves" and instead became absorbed into the movement of "the irregular flow of energy that defines this place and of which we are part." Similarly, Peter Ouspensky recounts that he felt himself being drawn to and becoming one with the waves: "It lasted an instant, perhaps less than an instant, but I entered into the waves and with them rushed with a howl at the ship." While scuba diving, Stephen Harrigan imagines himself becoming "disembodied and transparent, the seawater seeping through me as if I were some membraneous organism like a jellyfish." And during another dive, he explains the "rapture of the deep" as really being a rapture of belonging: "This was my territory. This was, in some sense, me. I was the ocean, and my body was nothing more than a particle within it, and uninhabited probe coasting through the radiant waters of the reef, receiving sense impressions." "I like to imagine being like water these days," reflects Brenda Peterson. "Call me pelagic," instructs Thomas Faber, insisting that by definition he too is "of or pertaining to the seas or oceans." Craig Lambert looks upon his role as a sailor in being a mediator linking the wind and the water through his rudder: "Suddenly, 'I' vanished and instead there was a unified field of water, wind, and a translator in their conversation; what is more, this translator was conscious, and hence could steer. The rest of the day was ecstasy."

Perhaps no one has expressed the sense of water dissolution of self more beautifully than the anthropologist Loren Eisley. Whereas Aldo Leopold, in a much quoted passage, admonishes us "to think like a mountain," Eisley instructs us to feel like a watershed: "Once in a lifetime, perhaps, one escapes the actual confines of the flesh. Once in a lifetime, if one is lucky, one so merges with sunlight and air and running water that whole eons, the eons that mountains and deserts know, might pass in a single afternoon without discomfort . . . Many years ago, in the course of some scientific investigations in a remote western county, I experienced, by chance, precisely the sort of curious absorption by water—the extension of shape by osmosis—at which I have been hinting. You have probably never experienced in yourself the meandering roots of a whole watershed or felt your outreached fingers touching, by some kind of clairvoyant extension, the brooks of snow-line glaciers at the same time that you were flowing toward the Gulf over the eroded debris of worn-down mountains."

For many, this is the key—the ability to recognize that the body is mostly water, and that perhaps it should behave like liquid. Byron Ricks, on an ocean kayak trip, feels his mood rising and falling with the waves as the tidal cycle becomes intuitive and internal, his blood responding to the moon's pull. Following his daily paddles, Chris Duff would engage in a simple ritual of connectivity: "I would feel the salt drying on my skin and offer a wordless prayer of thanks for this life. It was at times like this that I was aware how little I spoke, yet how filled my mind was with images and feelings. I breathed, ate, slept, and lived with the sea and the skies. Within that physical world of listening and patiently watching there was a growing spiritual connection that went beyond spoken words."

To become one with water is to set back the clock to the time of our beginning in that primordial, elemental sea from which we sprang. The longing for this temporal and corporeal union is perhaps the strongest of all feelings of hydrophilia. John Jerome believes that this is the explanation as to why we have such a strong desire to "immerse ourselves in a life-threatening foreign medium for which we, unlike other vertebrates, have no instinctive aptitude." To go into water, Jerome argues, is to be an intruder, to experience "the thrill of becoming the Other." By going back to water then, perhaps "we are trying to remember when we were not the Other," or as Michael Delp would have it, "end the pilgrimage where it began" by completing the circle. Those unable to reconstitute this umbilical connection to the past become severed from that which makes us human. In so doing, Brenda Peterson believes, these individuals become lonely by forgetting who they are and where they originated. "Does the sea make us so lonely because we've lost our connection and come keening to her shore like motherless children?" she wonders.

The pull of water upon the soul can be all-consuming, inducing a "flood of feeling" to the sensitive, as Jeff Wallach describes: "If only we could interpret rivers' messages. If only we could read the waters. If only we could ride the undercurrents to the end, and slide into a sea of emotion, headwaters of our humanity. If only we could give ourselves over to the flow and let it truly move us." What is it that the cetaceans and pinnipeds (the whales and the seals) recognized that made them leave the land and return to the sea? For Barry Lopez, the most intriguing question is the one that addresses why we humans didn't join them and instead remained behind, high and dry. "Along the very edge of these gravel bars are some of the earth's seams," Lopez writes. A place in which "a person with great courage and balance could slip between the water and the rock, the wet and the dry, and perhaps never come back." And as a provocative and emotion-filled afterword, Lopez concludes: "But I think it must take as much courage to stay." Is this, then, really the true difference between us and marine mammals: merely the degree of bravery that each respectively possesses?

At the end of his trip around Ireland by kayak, Chris Duff feels intense sadness as he must return to his terrestrial life once more: "The ache in my heart was the fear of losing the companionship of the winds and the swells, of losing the soul connection I had experienced on the ocean. I felt as though the ocean knew me better than any human could. I was unashamed to sing in its presence, to cringe in fear at its power, or to stare in utter wonder at its beauty." It is the fear of having to readapt to a dry ife in which he would have to be something different than who he had become on the sea, that scares Duff. In heartfelt prose similar to that voiced upon ending a romantic relationship, Duff contemplates the emptiness of a possible future existence: "What if I never again felt the power of a wall of green water rising and arching in fluid poetry on its final rush toward the perfection of sea meeting cliff? How could I live without the purity of those rhythms?"

### Water Reflections

What is it about water that intrigues us so, that alternately captivates and mystifies us, that ultimately inspires us? For centuries poets have directed their energies toward addressing such questions. The results, though undeniably beautiful, always seem to come up short. What water really is, and what the power is that it evokes to capture our souls, remains elusive. D. H. Lawrence, voicing such concerns, teasingly responded, in a vein frustratingly similar to Mallory's bathos about wanting to climb Everest simply "because its there," that it is the "third thing" that is important for understanding water. What then is this mysterious "third thing"? Here is how Lawrence explains it: "Water is $H_2O$, hydrogen two parts, oxygen one/But there is also a third thing, that makes it water/And nobody knows what that is." Modern nature writers continue to struggle to help identify this special essence of water.

Kathleen Dean Moore, describing the joys of having the hidden landscape suddenly revealed as a series of surprises when on a river journey, refers to water as "an agent of distortion and change, forcing a person to see things in new ways." Expressing similar feelings, Harry Middleton believes that the "chaos of raw possibilities" offered by water is as close as he can come to sensing true magic. Richard Nelson admits to being hypnotized and fascinated by breaking waves to the point of always being drawn to study them. "But still," he recollects, "I've only begun to learn. Perhaps there is too much difference between the human mind and the mind of water." For Jonathan Raban, water is the true "Primal Cause," functioning as a mirror to one's own existence. Craig Lambert believes that being suspended in a rowing shell in the transitional zone between the liquid and air "opens a window on mysteries hidden from those with solid ground beneath their feet," such that "sliding between dark and shadow, between sunlight and the obscure, is the region of discovery. Here the inchoate seeks from. Every area of creation has such a penumbra: venture capital, avant-garde arts, courtship. In such crucibles, imagination creates the future." Wallace Stegner's fascination with mountain rivers is an affair of the heart rather than of the mind: "By such a river it is impossible to believe that one will ever be tired or old. Every sense applauds it. Taste it, feel its chill on the teeth: it is purity absolute. Watch its racing current, its steady renewal of force: it is transient and eternal. And listen again to its sounds: get far enough away so that the noise of falling tons of water does not stun the ears, and hear how much is going on underneath—a whole symphony of smaller sounds, hiss and splash and gurgle, the small talk of side channels, the whisper of blown and scattered spray gathering itself and beginning to flow again, secret and irresistible, among the wet rocks." Others have engaged in similar riverine conversations, as, for example, when Eddy Harris recounts being able to "feel the spirit of this water rising up from the morning's mist" and hearing it "whisper" to him that he has nothing to fear; or at times of rain when "a river seems to speak to itself," spreading over the banks and "calling you in," as Michael Delp reflects.

For some, water serves as a metaphor for personal rebirth. Water aids a recovering Lorian Hemmingway: "I have felt the force of rivers deep at their heart. I have felt that force slam into my belly and carry me weightless into its dominion. I have dived into it naked, unafraid, watched snakes glide on its surface, seen it rise in fury tall as trees, watched waterspouts drink it deep into the funnel and then wilt before its vastness. I have seen it baptize, drown, and resurrect. It is the resurrection I want." For others, water serves as a means of return to the original birth. "Water is alive and in relationship with all of us who are blessed to be intimate with such a world-shaping, yet humble ally," writes Brenda Peterson. Reminding us that water is the first element we encounter while immersed in an "amniotic sea so rich that it recreates the primal ocean as we move again through all stages of evolution," Peterson pointedly asks, "However could we forget this first watery bond, even beached as we become in bodies that struggle against gravity, breathe high, harsh, and fast air, and at last lie down in solid ground?"

Yet, even with all these explanations, perhaps we must simply face the reality, as did D. H. Lawrence, in the elusive nature of what it is that makes water so very important to us. Should we really be surprised when, at the end, it may be impossible to put a rigidly defined label on an element known for its ability to resist containment by its physical nature? This is what Jeff Wallach thinks: "We cannot escape metaphors for waters because water itself is too large and encompassing to grasp. It slips through the fingers. Yet we are immersed, drenched, soaked, permeated by waters. Rivers are earth in flux, alive, something to fear and worry about and love with great awe. Like the writer Norman Maclean, we are all haunted by waters."

And finally, when all earthly explanations fail to capture what the true, special quality of water really is, some are forced to ascribe a spiritual nature to it. Michael Delp likens rivers to a divine presence: "Surely, rivers are the blood of the beings more capable and sensible than we are and to stand waiste deep in a trout stream is to make contact with the pure spirit of the gods themselves."

### Time Is But a Stream

Recounting Thoreau's famous maxim, Kathleen Dean Moore marvels at the sage's inferred ability to be able to visit or leave the flow of time at will: "This is an astounding idea, but it couldn't possibly be true. Time bowls me over, knocks me down, rolls rocks against my ankles. How could time treat Thoreau so gently and be so rough with me?" Thoreau, of course, was far from the first to liken the passage of life to that of water flowing in a stream. From our birth in the springs and seeps, through our adolescence in the turbulent headwater reaches, to our calmer maturity in the more placid lower reaches, and finally to our eventual death and dissolution in the sea, water carries us through life.

Modern writers ably follow in the tradition of Heraclitus, Thoreau, and others, by exploring the convenient metaphor of life's grand fluidity. Facing astern allows rowers, says Craig Lambert, to be compensated for the lack of an immediate future by a clear view of the immediate past, symbolized by the symmetry of the wake. He continues: "The finish of the rowing stroke, the release of blades from water, completes and begins a cycle: with the death of one impulse arises another. Release becomes recovery. I feather my blades and, on the inhale, glide softly up to the catch. The following sequence is an emblem of beauty and order. The stroke cycle and the life cycle are one." Joel Vance describes his enjoyment in sharing rivers with children for the first time. Let such rivers inspire you to take risks by allowing their "wildernesses to act on you," Jeff Wallach instructs. Ann Linnea paddles into the fury of Lake Superior to seek "a place of light, of understanding" in attempt to rescue her "heart from a twenty-one year silence." John Jerome makes the case that for his mother, "water, and immersion in it, was the organizing principle of her life." Bill Green goes further, toying at the edge of scientific assurance, with his insistence that for him, life itself is not only influenced by water, it is water: "And who are we that we should know these things? That out of the mingling of water and stone, out of the touch of sunlight, out of the carbon drawn in long chains, out of the mats of heme, the iron and manganese, the calcium and magnesium of ancient seas, the seas of our life's blood; that out of the helices and rings of matter, we should dream these dreams—this mysterious deep time we cannot fathom, only measure; these cycles of matter we cannot control, which pass through us, which link us irretrievably to all that is. 'Steep yourself in the sea of matter, bathe in its fiery waters,' Teilhard de Chardin said, 'for it is the source of your life.'"

In one of the most hauntingly beautiful passages ever written about water and lives, Michael Delp recounts how, after stealing his father's body from the morgue, he takes him back to their favorite fishing spot and buries him deep in a sandbar: "I waited until early evening, lit the lantern and then began dismantling the dam, only enough to let the water in, letting two logs drift away in the darkening current. The water sluiced over the dam, now inches underwater, over the stones, and sifted down, I am sure into my father's lips. I wanted to speak something to him in the dark but couldn't. He had wanted silence; wanted the sound of the river all around us." For Delp, this riverine burial is a way in which he can continue his relationship with his father each time he drifts over the spot in a boat: "I think of how his life comes back to me each time I fish, each time I step into the current. Mostly, I think of how both of us are carried by rivers, how his memory sifts through me like the current where only his bones are left to tell the story."

In a similar vein, Ann Linnea, standing on the coastal edge holding a dead friend's ashes, comes to peace with a four-year process of grieving: "I notice that my feet are underwater. The tide has come in to carry away that which is hers. I lift the cloth bag up in the air in front of me, kiss it gently, untie it, and slowly turn it upside down. Betty's remaining ashes scatter across the water's surface in a fine white line. The leading edge of a wave comes in and moves the ashes in around my feet in final salute. Then the wave recedes, carrying its precious cargo out to sea." Thomas Faber, too, longs for such an end, or a beginning: "When I die, please, scatter my ashes on the face of the waters. Warm waters, too, as I head off. Let me cycle and recycle in the tropics forever and ever . . . and don't mourn for me. I'll be in touch...when it rains. When it pours!" Tennessee Williams's dying wish, Charles Sprawson informs us, was to have his ashes scattered in the Bay of Mexico as close as possible to where Stephen Crane disappeared after a beautiful dive off the end of a boat, because "I've always admired the gentleman and I never had an opportunity to meet him." For Stephen Harrigan, a scuba dive into a dangerous cave offers a strange allure: "I saw how seductive it was, how death in an underwater cave would be for all its horror, an infinite form of regression. You would go back the way you came, swimming through a dark canal, divesting yourself of consciousness and matter until finally you were reabsorbed into the bloodstream of the earth." What was it that drew Virginia Woolf to waters to meet her end? And the image of Shelley, finally succumbing to his life-long morbid fascination with water and suicide, and calmly going down, still haunts.

### *Wet Dreams*

The art of drifting and dreaming over what Edwin Teale refers to as the "silken silence" of water has a long- established tradition. Michael Delp, suggesting that the reason might be either the slant of the reflecting light or the slant of his ranging imagination, easily falls into such reverie: "I drift between the literal world of the river in front of me and the other world of the river I have in my head." For Chris Duff, it is the mantra-inducing physicality of paddling that frees his mind to wander "back thousands of years or exploring some esoteric idea." Once when paddling, Duff encounters a feather upon the watersurface far from shore: "It stopped spinning, frozen for a second on its refection, then slowly turned again. Perfectly balanced, it floated with grace and beauty: a lesson of Zen on the open sea. Suddenly I didn't feel so alone." Ann Zwinger recounts how the water drops off her paddle blade to make spreading circles on the calm surface over which she floated, the pattern being likened to "the quiet breathing of the river, boundaries at once lucid and dissolving: quintessential riverness." At such times, Paul Theroux believes, one can enter into a trance and the experience become mystical.

As well as nurturing the body and soul, water has a way of fostering and encouraging intellectual pursuits. In those situations where the efforts of moving one's body over the surface of the water become secondary, the mind is free to wander far into realms seldom visited, plunging into depths rarely glimpsed. For Sigurd Olson, such an experience allows him the opportunity to ponder the intangibles of his life: "One day in the Far North, I paddled many miles down a great waterway. The wind was behind me, just a breath, but enough to make paddling easy and almost effortless. I watched vague islands gradually assume form, and points beginning to jut out from the mainland. I was alone with my thoughts, completely engrossed, and almost mesmerized with the idea of the unknown and the whole fascinating concept of mystery." Ann Linnea agrees: "When we deliberately leave the safety of the shore of our lives, we surrender to a mystery beyond our intent." On calm days, paddling for her becomes a meditation into the deepest part of her soul, in which she feels herself drawn "forward into understanding Mystery and Higher Purpose."

To reflect thus, with the gift of solitude, is a recurrent theme in the literature of water contemplation. Kevin Patterson, fleeing a spoilt relationship, attempts to find grounding in his life, paradoxically by leaving solid land far behind: "It wasn't at all clear to me that I was unhappy out there. I sat in the sun and stared at the water and watched the sails drawing the Sea Mouse along and thought over and over again of how enormous the sea was and how far I was from land and humanity. It remained beautiful to me. But less important for being alone." Fleeing a similar broken relationship, Paul Theroux wishes to completely disappear into the sea now that half of his life "had been eclipsed," his motivation being simple: "I paddled because it was a way of being alone."

To be successful at such mental tripping requires, of course, the ability to engage an active sense of wonder and allegory. Those who do this best, pursue the goal with a transcendental fervor to their reflections: "For at that moment I had a sense of unreality, of time out of mind— this clear water, this quiet river—and for that moment the Assabet became the idealized river of imagination, the way I had wished the river I grew up on to be and never was, flowing through an arcade of idealized trees, a river, just as Hawthorne said, to be found only in a poet's imagination—or a child's," writes Ann Zwinger, revisiting the landscape of Hawthorne, Emerson, and Thoreau.

# Part III
# Surfacing: Implementation

*Water must be a basic consideration in everything . . . We must restore the natural motion of our rivers . . . A river flowing naturally, with its bends, broads, and narrows, has the motion of blood in our arteries, with its inward rotation, tension and relaxation.*
> —Richard St. Barber Baker, *Man of Trees*, 1989

*Today's environmental problems are clearly recognizable as newly resurrected spiritual questions that have become matters of life and death for present-day humanity. They cry out loudly, demanding solution after so many centuries, solution with new human capabilities. The consciousness of humanity as a whole has completed its descent into earth and the kingdom of dead laws. Now it has become the obligation of the individual—the "needle's eye" of the human race—to travel the road to the realm of life, to a rebirth learned from water's being . . . We have become water's destiny, and from this point onward water becomes ours.*
> —Theodore Schwenk, *Water, the Element of Life*, 1989

*All truly sustainable, and therefore successful, environmental restoration projects are as much about restoring degraded human—nature relationships as they are about simply restoring degraded physical landscapes.*
> —Robert France, "(Stormwater) Leaving Las Vegas,"
> *Landscape Architecture Magazine*, 8/2001

# Reversing the Flow:
# Human—Water Relations

### *Written on the Water*

Water transcends boundaries—professional ones as
well as physical ones—with ease, and continues to be
an inspirational force in the diverse field of environmental letters. Though many books attest to this,
there is a new movement directed toward exploring
different literary approaches for restoring human—
water relationships. These works purposely strive to
fill the niche between the technical but dry prose of
scientists and land managers, and the personal but
often nonfactual prose of nature writers. Of these
crossover books, several are particularly memorable
for experimenting with fresh ways of immersing
human consciousness in watersheds. All describe a
sense of place in which writers are not mere visitors
but are instead inhabitants. This is, of course, where
Thoreau, the master of the microcosm, reigned
supreme.

In *Waterstained Landscapes: Seeing and Shaping*

*Regionally Distinctive Places*, Joan Woodward develops a novel form of describing how water simultaneously shapes both nature and lives. A fictitious character arrives in the Denver area with fresh eyes and undertakes a literary equivalent of a geographic information system, or GIS, analysis of site conditions. Her protagonist ranges over the western landscape marveling at the overt traces of water's passage upon the desert, and begins a journal of her attempts to sift through deeper layers. Increasingly preoccupied by searching for knowledge of the local, the heroine alternates field excursions with library explorations. As the character completes her investigation of the landscape and the water stains upon it, the "novel" closes with her departure, the reader being left with the impression that in a society of gypsy-scholars, the tools of ecological history and inquiry are easily transferable from one locale to another in order to begin searching for what Chris Childs has called the "secret knowledge of water."

Anuradha Mathur and Dilip da Cunha's *Mississippi Floods: Designing a Shifting Landscape* attempts to portray the dynamic nature of the significant river in the light of the ensuing conflicts of humans battling against and attempting to control the environment. The massive damage resulting from the 1993 floods, and the continued lack of wisdom of humans who persist in living in blind ignorance within the floodplain, led Mathur and da Cunha to undertake a detailed study of the representation of the river through time. Information from hydrologic models, old stream course maps, scientific reports, drawings, photographs and paintings were integrated into a series of layered silk screen prints that combine history and engineering in order to securely place humans within the shifting riverscape. This intriguing fusion of art and science attempts to show that by endeavoring to capture a dynamic river within controlled levees, we inevitably lose some of our own naturalness in the process. In the end, floods are really life-giving, not taking, events that should be understood and cherished.

From 1937 to 1974, the "Rivers of America" series of sixty-five books was published, each one focusing on the historical human development (what today we might call "exploitation") of a particular river of note. Recently, in *Lewis Creek Lost and Found*, Kevin Dann provides a regional portrait of a watershed as seen through the eyes of three nineteenth-century explorers of the landscape; one, a historical writer, the other two, scientists. Through these accounts, centered around a single watershed, Dann presents voices that explore the relationship between water nature and water culture.

David Cassuto's *Cold Running River*, Valerie Rapp's *What the River Reveals*, Freeman House's *Totem Salmon: Life Lessons from Another Species*, and my forthcoming *Still Waters, Lost . . . Still, Waters Lost: An Environmental Narrative of Sin and Purgatory* represent a new genre of what might be referred to as watershed recovery. These works present a historical survey of the life and times of rivers: how the systems functioned before European contact, accounts of "discovery" and early settlement, the gradual degradation resulting from poor land management and collapsing fisheries, and finally the gradual recovery through watershed stewardship and restoration. These watershed biographies recount the frustrations and celebrate the hope for a future in which the rivers can be reconnected in a sustainable way to both their watersheds and to the hearts of the human inhabitants who reside there.

### Healing Nature, Repairing Relationships

For much of its history, the environmental movement has directed its attention to the preservation of undamaged landscapes. Increasingly recognized, however, is the disturbing fact that we live in a "world of wounds." Many are now examining previously held beliefs in an assumed dichotomy between "nature" and "culture." These individuals are instead turning to the practice of ecological restoration as the means for reestablishing a close working relationship with the environment. The act of restoring, remediating—in other words, healing—degraded water is an act of reciprocity, important not only for improving the quality of the outside environment of nature, but also that of the internal environment of the psyche, or human nature. For example, the Native American healing center created in Arcata, California by Laura Kadlecik and Mike Wilson, builds upon the mutuality of restoring a degraded physical environment at the same time as restoring a degraded spiritual environment. The creation of a wetland that not only functionally treats stormwater on-site but also aesthetically serves as the focus for ritual healing practices, goes far toward integrating the worlds of the inner and outer.

Being likened to gardening, ecological restoration can provide a model for reconnecting people to the world in a healthy relationship. By acknowledging the human presence in the environment, and accrediting past site conditions and the role of humans in shaping them, restorationists provide a means for reimagining a future in which humans do not sit idly by on the sidelines as the world goes to ruin, but instead engage with nature in a healing partnership. In a society that has lost much of its sense of ritual and honor, restoration provides a vehicle for participants to experience nature in an expressive act. Importantly, restoration is a public participatory process rooted in the wisdom of understanding the site conditions in the pristine past, the degraded present, and the rehabilitated future. The restoration of the Las Vegas Wash and creation of the ensuing Clark County Wetlands Park, for example, benefited immensely from the help of thousands of local individuals and business visitors in a massive, Cecil B. DeMille–like orchestration of marshaled energy and effort.

The key to this process is the fact that through ecological restoration, people develop a way of increasing the bonds of care with which they experience the world about them. This occurs for the broken land and water that is repaired as well as those areas that are non-degraded and therefore presently unneeding of restoration. The restoration of a salt marsh in San Francisco in which locals participated is one example how this may develop. Faced with a need to have a large public gathering place for the millennium celebrations, city officials selected the restored marsh. However, the outcry from local neighbors and the concerned public—the very individuals who had themselves spent long hours in helping to replant the salt grass—was such that officials were forced to select an alternative, dry location that would not damage any wetland.

Direct public engagement in the physical act of ecological restoration is therefore the route to success. Nowhere is this better demonstrated than in the restoration of the landscape and waterways of Sudbury, Ontario, a region devastated by a century of acidic deposition originating from the world's largest single source of sulfur dioxide. The "trout release program" described by John Gunn was a remarkable event wherein hundreds of individuals, including unemployed miners and their children, gathered in a parking lot and were given a map of the regional headwaters in addition to a plastic bag containing a small fish. By the time the cars had arrived at their respective release locations, the fish often had been named and had become adopted members of each family. The follow-up press conference was equally useful and amusing. In response to the Mayor's offhand comment that "Wouldn't it be great to be able to drink the waters like in the old days," Gunn looked purposely at the rolling cameras and placed his arm around the Mayor, challenging him: "Why don't you and I in two years time come back to this creek and drink the water together? Why don't we commit to that?" Afterwards, the panicking mine owners, pressured by the Mayor's office, asked Gunn privately if any of the future intended release sites were to be near their outfalls. Of course the answer was "yes." Two years later, however, mining officials called Gunn several days before the publicized release event to insist that the fish would in fact be let go at their mine sites and that cameras would be present to film the fish as they swam their way down the streams, newly restored following the expenditure of several million of dollars. It is no wonder that this success story was the winner of the Earth Summit Award at the 1992 Rio Conference.

In some respects, ecological restoration can be thought of as a human gift back to nature. As William Jordan states: "The restored ecosystem is something we can offer nature in return for what nature has given us, or what, if you prefer, we have taken from it." By uniting human and natural communities at the local level, restoration is a way for participants to deeply immerse themselves in a positive act of benefit to both communities. Restoration in this sense is a positive process. Rather than always negatively reacting to bad news by moping about what seems to be increasingly wrong with the world, restorationists can proactively and positively go out and do something beneficial about it. Only a year after its opening, the London Wetlands Centre—an urban ecotourism park developed from a series of abandoned Victorian reservoirs—already has a long waiting list of volunteers eager to continue an ongoing working relationship with the restored water brought back to life.

We need to redefine how we regard water, for, as Jim Patchett and Gerry Wilhelm state, we have become so detached from the natural world that we have become unmindful of the mutuality of our relationships with it. We take for granted or have simply forgotten how much our lives are enriched by water. We need to find models to help us remember.

One solution might be the sort of restoration ethic that Bruce Hull and David Robertson argue for, one that "instills respect in us for the land. It builds into our culture an appreciation and respect not just for nature, but for our relationship with nature, and not just for wild nature, but for all forms of nature from parks to parking lots."

In the end, restoration is as much about healing the sinners as it is the sinned against. By addressing the sins of our predecessors, we accept our obligation to nature. Restoration should therefore be approached as an act of contrition, an acknowledgment of guilt, a restitution for what caused it, an expression of remorse, and as a dedicated effort for renewal. For as Peter Forbes so aptly states "The path to the great remembering is through the healing of land conservation and the healing of ourselves, through a million different ways to show our forbearance to reconnect with the life that is around us."

### Day-for-Night Resurrections

It is really the most egregious injury we can ever inflict upon living nature. It might even be evil, at least at the beginning. At the end, the situation is hardly better, for it is here where, to use the old adage, insult is added to injury. "Insult" in the form of forgetfulness. For what can be worse than to not only physically disappear from the world of the living, but then also to conceptually vanish from the world of the memory. In the brilliant and hauntingly disturbing Danish film, *The Vanishing*, the protagonist becomes obsessed with keeping alive the memory of his girlfriend who has been abducted. Then, in one of the most horrific endings in all cinema, he finally determines the fate of his beloved. Memory is a key to giving meaning to life, be it the life of an individual or the life of an ecological system such as a river. For it is memory and its relationship with love that is the "ultimate reason," as Thorton Wilder concludes in *The Bridge of San Luis Rey*.

It is somewhat ironic that Wilder uses the metaphor of a bridge to span between the worlds of the living and the dead. For it is with construction of the first such physical bridge, then the subsequent widening of it, then the building of another bridge nearby, then the widening of that one, then the joining together of the two as a linear parcel of "reclaimed land," that allows urban rivers to incrementally disappear from our lives and hearts, and with time, eventually from even our collective consciousness.

Streams have not fared well with urbanization. Fears about flooding proselytized by hydro-engineers led to watercourses becoming straightened. Stormwater runoff courses off impervious surfaces, dumping a cocktail of contaminants directly into the water. Constrained within concrete channels or placed in open pipes, it is only a matter of time before the rivers become completely covered and eventually forgotten.

Today a new phenomenon is revitalizing urban renewal projects directed toward the unearthing or "daylighting" of buried steams. Daylighting is thought to be one of the most radical expressions of a new consciousness of interaction with surface waters. Today, close to three miles of waterways have been liberated from their prisons in more than twenty projects across the United States, including the pioneer efforts by Wolfe Mason Associates in Berkeley, California. And in Zurich, where the memory of flowing streams is even foggier due to a greater accumulation of urbanized time, fully nine miles of streams have been restored in this fashion, awakening the populace to the forgotten wonders of water.

Notwithstanding the technical obstacles that must be addressed when daylighting urban streams, the major challenge is often a sociological one. For sad as it is to take a dynamic living being and imprison it within the ground, it is all the sadder still when the memory of that living being is lost. The result is that urbanites often fear open water: newly uncovered streams will flood, vermin will breed, children will drown, undesirables will flock to the neighborhood, etc. Because of this, daylighting projects frequently need considerable attention devoted to educating the public about those elements of joy, adventure, contact, and contemplation that rivers can provide.

Enlivening rivers is nothing short of enlivening life itself. Daylighting is the ultimate manifestation of what might be called "resurrection ecology." By bearing witness to the return of a riverine ghost from the past, we develop a new respect for water. Like a new pair of

lovers meeting for the first time, or several old friends becoming reac-
quainted after a long estrangement, the tentative approach to the new
partner is one of initial excitement and adventure, leading to a more
mature joy, then of intimate, exploring contact with the other, leading
to reflective contemplation about one's own self. For it is through gaz-
ing at surface water that the observer, just as much as the observed,
becomes reborn. And, as in all relationships, the observer sees himself
or herself through the mirror of the other, the beloved, in this case the
new river.

*And so water becomes the great teacher at the moment when abstract con-*
*sciousness crosses the threshold to that other consciousness that once again*
*befriends itself with laws of life.*
　　　　—Theodor Schwenk, *Water, the Element of Life*, 1989


*We went out to the pump-house... As the cold water gushed forth, filling*
*the mug, I spelled "w-a-t-e-r" in Helen's free hand. The word coming so*
*close to the sensation of cold water rushing over her hand seemed to startle*
*her. She dropped the mug and stood as one transfixed. A new light came*
*into her face. She spelled "water" several times... suddenly turning round*
*she asked for my name. I spelled "Teacher."*
　　　　—Ann Sullivan, letter *in* Helen Keller, *The Story of My*
　　　　*Life*, 1887


*Water is a sublime teacher . . . better understanding of water will give each*
*of us a way of seeing our self and help lead us to a better life.*
　　　　—William Marks, *The Holy Order of Water: Healing*
　　　　*Earth's Water and Ourselves*, 2001

# Turning the Tide:
# Rehydration Education

## *Liquid Learning*

There is a crucial need to connect people to their landscapes, to anchor them in a deep understanding of the important role that water plays in their lives. Conferences like the recent "Water Sensitive Ecological Planning and Design" and "The Role of Water in History and Development," although undeniably important, offer instruction without engagement. We need to search for non academic, less dry ways in which to educate about water. In other words, we need to encourage the uninitiated to experience water. And thereby, to love water.

One program of noted success in reconnecting people to water, is the International Rivers Network's "River of Words," or ROW. The organization's vision is to "seek a world in which rivers and their watersheds are valued as living systems and are protected and nurtured for the benefit of the human and biological communities that depend on them." The means to effect that goal is by developing a worldwide understanding of how rivers sustain ecological integrity, social justice, and human rights. By cosponsoring (along with the Library of Congress and the United States Poet Laureate for 1995-1997, Robert Hass) an international contest for children, ROW focuses on encouraging youths to learn about their watersheds through environmental poetry and art. By engaging such creative imagination, ROW hopes to foster responsibility for the waters that inspired the projects. Every year, the award winning children and their parents meet in Washington for a public reading and exhibition. Through allowing children to explore the simple beauty of place through art and poetry, links are revealed "between people and nature, the physical and the spiritual." As Robert Hass states: "We need both things—a living knowledge of the land and a live imagination of it and our place in it—if we are going to preserve it. Good science and a vital art and, in the long run, wisdom." There is now a "water drop" patch for Girl Scouts to earn upon entering the ROW contest and exploring their own watershed.

Water can be just as fascinating to imaginative adults as it is to searching children. Both can be a sponge for water's wisdom. With this in mind, Elizabeth Cavicchi has for several years taught a course at Harvard University entitled "Exploring Water Through Ways of Doing Art and Physics." Her justification for undertaking such a course was simple: "Water is delighting, surprising, intriguing. All of how water engages us—for play and splashing; with spontaneity, reflectiveness, wonder—are openings to teaching and learning with water . . . For in responding to water and each other, our ways of thinking and interacting are also changing and developing. And water's changingness and fluidity evoke responses from us that encompass both artistic expressions and physical questioning. Art and physics, by being combined together, continually infuse our explorations of water with more to notice and try." To complement the assigned readings, the course contains a laboratory component in which students experiment with water flow forms, hoses and sprinklers, drip and ripple tanks, cups and tubing, etc., using photography and art to capture the nuances of how the liquid behaves.

If one goal of education is the exploration of the unknown, nowhere is this more important than in the exploration of historical infrastructures that have become all but lost from our urban consciousness. In John Stilgoe's book, *Outside Lies Magic: Regaining History and Awareness in Everyday Places*, the protagonist, referred to as simply the "explorer," uncovers culturally layered landscapes. The word "magic" here recalls the wonderful and often-quoted precept of Loren Eisley that "if there is magic on this planet, it is contained in water." The Surfrider Foundation has produced a video called "From Sea to Summit" that educates about the hydrological cycle and urban runoff and non-point source pollution through means of famous snowboarders, skateboarders, and of course, surfers moving their way through the watershed.

By conceptually uncovering our past waterworks, we are brought to a deeper understanding of how much the lives of our cities have been, and currently are, structured by water. Through use of innovative computer simulations that take the explorer on a tour within old watermains in Boston's Back Bay Fens, or by generation of an interactive website whereby explorers can sequentially peel away cultural and physical layers of Rome's complex hydrology as if participating in an archeological dig, educators such as Kathy Poole and Katherine Rinne are developing a theater of dynamic evolution that reinforces the fluidity of both water and history. In the case of the Back Bay Fens, computer models address the long relationship of humans to water through use of infrastructure systems for stormwater control, sewage treatment, and recreation. In the case of the 2,800-year history of Rome, the focus is on how the complicated and layered water system infrastructure has impacted and shaped the urban social life of the great city as well as defining its unique physical character. These two electronic explorations, as well as the less detailed but more general *Aqua Venturer* CD-ROM from the Water Environment Federation, share in common a belief in the power of history to shape the future: that by witnessing our aqueous past, we are brought toward cherishing and protecting our liquid present, and therefore preserving our wet future.

### Visible Water

Ivan Illich pessimistically contends that "the city child has no opportunity to come in touch with living water. Water can no more be observed; it can only be imagined by reflecting on an occasional drop or a humble puddle." For many urbanites, knowledge of water really extends no further than watching it fall from the faucet to the drain, or giving a desultory last look at it and our bodily wastes before they are flushed away. On the global stage, water, if it is discussed at all, is done so with regard to its role as a resource of national pride or envy. How can people be reminded that water is much more than an inert receptacle of our waste or something to pour into a glass in which we sprinkle flavored crystals before consumption? (recently the Olive Garden restaurant chain has started a campaign called "H₂NO" to discourage people from even drinking water in lieu of other beverages) And how can the population be mobilized to protect water from the continued and increasingly more serious threats posed to its ecological integrity?

The first stage in the process of conceptually and physically revitalizing water must develop through an increased awareness of its presence in our environments and in our lives. This is critical, for one does not protect what one does not love and respect, and one does not love and respect what one does not first recognize. And sadly, as a society, we have become for the most part collectively blind to water. Today, fortunately, there is an active group of imaginative individuals, landscape architects and artists, who are dedicated to celebrating the wonder and joy of water in their designs. These works reveal to the public the playful magic of liquid by blending art and science, and go far toward awakening respect for the element through fostering intimate observation and sometimes even direct contact.

The Don River in Toronto is, like many urban rivers, a polluted and mostly forgotten system whose riparian floodplain today functions as a corridor for automobiles as it did once for wildlife. Recently, a network of trails and observation spots was constructed to bring pedestrians and cyclists to where several two-story, white plastic molars, designed by Noel Harding and referred to as "Elevated Wetlands," stand prominently at the water's edge. Each giant tooth is topped by a small solar collector powering an internal pump that sucks up a small amount of water from the river and distributes it over mini-wetlands that have been created in the crown of each tooth. The polluted water then percolates down through a cleansing "soil" consisting of recycled plastic containers and the roots of toxin-removing plants, exits the molar, and returns to the Don River free of significant contaminants. Although only a minuscule volume of the river's water is actually purified by this

means, the art project successfully reveals and instructs about the ecological, kidney-like, purifying function of wetlands. Experiencing wetlands in this fun, silly, and admittedly artificial way encourages participants to preserve the few remaining natural wetlands in the valley. As
Peter North writes, the project was "conceived as a notion of isolating
or 'elevating' a natural process of a wetland in order to gain awareness
and provide commentary on our distanced relationship from the environment."

In Chengdu, China, with one important exception, the Fu and Nan
Rivers race by the rapidly developing city, unknown, unloved, and certainly unprotected by inhabitants focused on modernizing their lives.
Located along the banks is an installation called the "Living Water
Garden," a two-hectare riverfront park in the shape of a fish, the symbol of regeneration in Chinese culture, through which river water is
diverted and treated in a series of sculpted wetland cells. Located at the
tail of the giant fish is a clean water fountain designed to resemble a
chambered nautilus that celebrates the cleansed water and supplies a
splash pond where children are drawn like a magnet. The inspiration of
this project, like that of the Elevated Wetlands, was to raise public consciousness. For as Betsy Damon and Anne Mavor of the "Keepers of
the Waters" organization state: "If we want to continue life on this planet, it is time to create a new culture that understands and values water's
true nature. It is not enough to build isolated projects; we have to change
the culture so that future generations grow up with an intimate relationship with water. In this new culture, ethical decisions about water would
be second nature. Instead of a constant struggle between interests,
aligning economic and ecological health would become easy."

When visited in its pristine state as remote lakes and streams, we
marvel at the beauty of water: "What a singular element is this water!"
exclaimed Thoreau in about as excited prose as he ever mustered. When
encountered in our urban settings as stormwater runoff, we retreat from
it in fear and loathing: "Filth of all hues and odour seem to tell, what
street they sail'd from, by their sight and smell," recounted Jonathan
Swift. It needn't be these two extremes, however. Although Loren
Eisley's famous maxim that "if there is magic on this planet, it is contained in water" is widely quoted, few realize that he was actually referring to stormwater puddles on top of an urban roof. It is possible there-

fore to regard water in our developed landscapes as a design resource rather than as a waste product. No one has accomplished this more convincingly than Herbert Dreiseitl. In his own words: "The way water is handled in towns shows more than the mere technical ingenuity of its citizens, it reflects myth and religion and shows the spiritual constitution of people living in a water culture."

The question "Which of us could say that we have experienced water sensually in all its dimensions to say nothing of understanding it intellectually?" underlies all Dreiseitl's work. Several of his projects in Germany illustrate the visionary ways in which water can be appreciated and revered as an element of beauty while at the same time it is actually improved by the completed design. In a typical project, water enters the system through rooftop rain gardens that remove contaminants before sending the water into basement storage cisterns. The water is then recirculated upward through solar-powered pumps to where it is released inside the building along a set of stained glass waterfalls, the descent of the water sucking cool outside air into the building through a network of openings as a sustainable air-conditioning system. This water is then sent outside into a stormwater-reflecting pool where visitors are encouraged to directly experience the water by immersing themselves in it. Finally this water is moved back inside the building to be used as a gray-water source for the toilets. Other projects focus attention on stormwater, not as a nuisance, but as an art form. Rather than hide the water in roof downspouts that empty straight into underground catch basins, sputters at a hospital dramatically shoot the rain off into space as a cascade that falls into cleansing swales, the sound of the ephemeral waterfalls providing therapeutic solace to the patients inside. For in the end, as Dreiseitl so aptly explains, "To do justice to water, we have to go into the waterworld ourselves, experiment with it and learn to think in an integrated and interdisciplinary way about its flow and flexibility." In contrast, therefore, to many fountain designers whose work focuses on entertainment, Dreiseitl's projects are significant in that they are also didactic features serving to reconnect fountain water to its source in the larger watershed.

For environmental sculptor Basia Irland, water represents the slippery template upon which, paradoxically, we can anchor ourselves in a shifting world. Interested in the "symbolic, metaphoric, magical, literal, political, social, biological, botanical and ecological nature of water," her work both educates and connects, and has been referred to as "an extended investigation of and hymn to water." One of her projects is a collection of portable sculptures that house research, maps, water samples, logbooks, hydrologic reports, photographs, and video documentaries in a series of books of nature, referred to as "hydrolibros." Another project, "Desert Fountain," instructs residents of Albuquerque about the importance of water to their community. Interestingly, this sculpture only flows when a sufficient amount of rainwater or snowmelt can be collected on the museum roof and then piped to the installation site, where it cascades over a series of outstretched hands. It is during the dry periods, however, that the fountain truly functions. For at those times, the bronze work of art with its empty hands offers a silent testimonial to the ephemeral nature and therefore preciousness of water in such a desert environment.

The "Mystic River Journey" in Sommerville, MA represents an important hybrid of environmental education and public art. Here, in a location where the Mystic River is cut off from the city and mostly forgotten due to an interstate highway, large and colorful panels have been assembled into an eighty-foot long mural and mounted at an important intersection. What is most interesting about the project, Lisa Brukilacchio and Jennifer Hill explain, is that its subject arises through transforming the experiences of urban youths developed during canoe trips on the nearby river into a physical work of art that is progressively being expanded. Each year, students working under the direction of resident artist David Fichter tackle different themes of the river's history and biology, all motivated by a desire to increase the presence of the river in the lives of not only the participants themselves but also those of the thousands who drive by the installation site daily.

Given the developing local awareness of the Mystic River and the growing concerns about ongoing environmental threats, the Sommerville Arts Council brought together artists and thousands of people over a weekend in a celebration called "ArtBeat and the Windows Art Project." Local storefront windows were turned into mini-exhibits about Mystic River issues such as non-point source pollution and open space protection. According to Lisa Brukilacchio and Jennifer Hill, one of the most popular and successful events in terms of public outreach

was the "What does the Mystic mean to you?" activity in which participants were asked to pencil their answers in chalk on the middle of the
cordoned-off street: "Serious responses were matched with whimsical
answers and youngster's drawings, but the act of making a statement in
a public space was clearly one of empowerment, bringing a larger public into the dialogue of what a river means in an urban context."

## Wet Intentions

Water has always been a rich source of inspiration for artists. One
goal of artists is to reveal the novelty in the commonplace. Through
temporal confounding, artists can remind us of the past, alert us to the
present, and guide us toward a more positive future. One way to effect
such an outcome is through the performing arts.

Performing arts were born as a bridge between the natural and
cultural worlds. Early cultures, more closely embedded in nature than
those of our own, engaged in ritualistic acts developed from close
observation with the world about them. Today, some who feel themselves cast adrift in a faithless society, strive toward finding a means of
reintegrating and re-experiencing nature through performance arts.
Many of these activities are centered on water, for as Mark
Dannenhauser states "Water is not only an important integrator of environmental disturbance, it is also an important integrator of environmental restoration" that interlinks the artistic community, ecological
planning and design, and local landscapes.

"Common Ground" is a British organization dedicated to linking
culture with nature by enabling communities to rediscover their local
environment. One particularly interesting and successful endeavor is
called "Confluence," which attempts to unite watershed residents to
their rivers with music. Inspirations for the new works (from a watershed composer in residence) come from the river itself: locations, local
stories, legends, and myths. In 2000, one watershed organized "The
Water Market," a day-long celebration about all things wet, patterned
after a medieval village festival.

One New Mexico town recently developed an innovative way to imple-
ment the River of Words (ROW) message. A parade of children winds
down the main street carrying banners covered with aquatic artwork.
And, most interesting, shops in the town have a little basket next to their

104

tills where patrons are invited to remove or deposit a riverine poem.

Environmental artists bring to light connections between culture and nature that often go unrecognized by specialists. As Mags Harries and Lajos Heder explain: "One of our major roles is to create new contemporary understandings, images and metaphors for the relationship of water and communities. These new metaphors can bring new and much needed energy to water related design and the renewal of communities." Harries and Heder elaborate that as artists they can "also play a useful role in interpreting the function and role of water-related utilities in the community . . . [by creating] fresh encounters and perceptions that lead to real understanding, learning and enjoyment." Artists therefore function as teachers, choreographers, narrators, and above all, as integrators, bringing people and places together.

Performing arts instill a sense of common unity or community among participants in the shared experience. By amplifying image, metaphor, and meaning in which delight is the key, the public can engage with the environment in a direct and highly personable manner at the scale of the local.

In a performance entitled "Re-membering the Muddy River," a group of Boston residents were motivated by a desire to restore the memory of a particular region of a river that had been buried and covered by a parking lot and road. Over the course of a day, this group of "reclamation artists" ferried buckets of water from one side of the open river to the other, leaving a trail of sprinkled blue chalk along their route, symbolizing the forgotten river course beneath them. Initially somewhat timid in their efforts, as the day progressed along with the strength of their convictions, the emboldened group began to purposely parade into the busy traffic as a guerrilla act, educating motorists that they were really driving atop borrowed land. Due to such awareness provocation, plans now exist to daylight or open up this section of the river.

The "Bronx River Golden Ball" focused on reconnecting people to a degraded river that had been all but forgotten. In a "desire to put our bodies where our metaphors were," Mags Harries and Lajos Heder engaged a flotilla of dozens of canoes that floated a giant, gold-leaf ball ten miles down the river while being accompanied by a group of dancers performing an extended performance piece along the shoreline, and thousands of people attending to engage in cleanups. Local excitement over rescuing the river from oblivion led to a repeat performance at which the governor participated (in a canoe, not dancing) and then later announced a grant of eight million dollars for river revitalization and streamside green space development.

The Earth Day celebrations in Concord, Massachusetts, are of a scale that one might hope for, given the town's historic significance as the birthplace of American environmentalism. Two performance elements demonstrate the means by which the town, located where the Assabet and Sudbury Rivers join to form the Concord River, attempts to recover its riverine past. Organized by the "Musquetaquid Group," named after the Native American name for the region's waters (the place where water flows through the grasses) the "River Parade" consists of giant puppets of river creatures, some requiring up to half a dozen people to carry, winding their way through the town in a Carnival-like celebration of the aquatic world. Before the joyous parade, in a more solemn ceremony, early morning participants carefully offer their personal floats made of natural bric-a-brac collected from the neighborhood, to the river in homage. By releasing their floats to the water at the precise site where the rivers intersect, the modern participants are themselves intersecting with an earlier time when such locations in the landscape were given the attention and respect that they truly deserved.

Another project, "A Gathering of Waters," was designed to reconnect diverse communities along the complete 1,875-mile length of the Rio Grande. A special canteen, the "River Vessel," was passed downstream from one community to the next, each contributing water from their own locale to the integrated mixture. Over a period of five years, the canteen and accompanying logbook traveled by boat, raft, canoe, automobile, horseback, mail, and runners in a relay, each group or individual handing it off to the next in line on the downstream journey. As the canteen passed through Pueblo villages, the tribal elders offered blessings and performed ceremonies about the importance of the great river. Basia Irland has stated that her aim was to "restore symbolically a natural function of the river and generate enthusiasm, and a sense of continuity" that celebrated the great river and its cultures. Significantly, the entire project was inspired by the sad reality that due to many diversions and much overuse, the actual river now fails to complete its natural flow to the sea. The final release of the gathered waters of the Rio

106

Grande from the canteen into the Gulf of Mexico was therefore a huge celebration designed as an act of compassion and as a gift.

Since 1996, the spring-migrating herring making their way up the Mystic River in Massachusetts have been accompanied along their last 10-kilometers by the presence of hundreds of runners participating in the annual "Herring Run 10K Road Race & Mystic River Celebration." Over the years the event has grown in popularity and has brought attention to both the recreational potential of this often ignored river as well as the plight of it's fish. At the dam that serves as the final, unbreachable barrier to the migrating fish, a carnival atmosphere occurs where, as Lisa Brukilacchio and Jennifer Hill state, "educational and entertainment activities [such as display tables, puppet shows, music, fish paintings, etc.] occupy the runners and visitors while they wait for race times to be posted." And along with awards being given for the fastest race times, the local watershed group presents their annual prizes for leadership in education and in advocacy. Meanwhile, admix all the joy, it is impossible to ignore the sad plight of the thousands of alewife and herring that pile up at the dam in a futile attempt to get up into the headwater lakes to spawn, instead falling prey to the many birds that gather around. The good news is that the recurring race and celebration has focused enough attention on this issue that plans are being drawn up to rebuild the deteriorating dam and to install fish ladders.

If, of all the elements, it is water that has the premier ability to capture minds, hearts, bodies, and souls throughout the ages, then certainly it has to be admitted that fire must be regarded as being a close second. Strange, then, that it was not until 1994 that someone had the wonderful idea to combine the two elements as a dynamic art installation project directed toward fostering community and urban renewal. Quite simply, there is no water art project anywhere that has captured the public imagination to the same degree as that of Barnaby Evan's internationally renowned, award winning, and immensely popular "WaterFire" installation along the Waanasquatuchet and Moshassuck Rivers in downtown Providence, Rhode Island.

WaterFire consists of one hundred bonfires that emerge from the water along two-thirds of a mile of river canals. The area is bordered by an urban "water place" park and promenade located within a newly revitalized downtown core of high-rise office buildings. At dusk, black-

107

clad volunteer "performers" (there is an enormous waiting list) in boats (named after gods and heroes from Greek mythology) shuttle back and forth among the string of braziers as they tend the flames. The evocative aroma and radiating heat from the burning cedar and pine embrace the senses at the same time as the eyes delight in the shadows cast by the flames flickering across the dark surface of the water flowing silently beneath the pyres, while the ears savor the beautiful music—a wonderful and eclectic mix of classical, Gregorian chants, and gothic rock—piped in through a network of speakers built into the walls of the riverside promenade. Designed to transform our perception of urban water, Evan's installation intentionally immerses us through synesthesia, where the experience of one sense results from the stimulation of another, the integrated whole being much greater than the sum of the separate parts. The public experiences the urban rivers by strolling through the art installation across water-spanning walkways and elevated bridges as they become enchanted by the fire, aroma, music, and light dancing upon the water.

WaterFire is a kind of primal tribal ritual recalling a Viking burial or the cremation ceremonies along the Ganges, and it has been instrumental in the modern renaissance of the city of Providence, allowing guests to discover and residents to rediscover the central importance that water can play in their lives. Although initially intended as a single event for the 1994 First Night celebration, the entire community became so engaged in the product and process that today a nonprofit group exists that arranges a dozen or so annual "performances" to which over 300,000 attend. As the web-page intones: "The broad support for WaterFire Providence and its power to attract millions of visitors are eloquent testimony to the power of public art and to its capacity to restore our urban and social landscapes."

The accolades from the general public and press about WaterFire read like a litany of hyperbole: "an uplifting experience," "compelling and enigmatic," "overwhelming impact," "engages with emotional power like very few other works of art," "one of the loveliest spiritual/community experiences of my adult life," "accomplishes what many other public installations only aspire to . . . we are pulled out of as well as into ourselves," "without a doubt the greatest participating artist event since the ancient pagan worshipping times; truly a spiritual and magical event—romantic and beautiful," "the most moving public art piece I've ever experienced," "the best work of public art or sculpture/perform-ance I have seen anywhere in the world," "the people stood transfixed on the banks," "THE most wonderful event I have ever attended in the USA," "has the power to transport us to another realm, and in the process of that transport, we are transformed into more appreciative, more pleasant individuals," and if all that were not enough: "I was total-ly immersed in the piety of the moment. The visual effect and the ambiance of the work will be embedded into the visceral memories I will carry with me my entire life," and finally, "thank you, thank you for so much beauty; I listened to the violin echoing under the bridge and if it had been my last experience on earth it would have been perfect."

*And water is wisdom's very element, the focus in which wisdom is concentrated and out of which its activity flows into every least and greatest living thing. Indeed, it is because life is wisdom and water wisdom's element that there can be such a thing as the water of life.*
—Theodor Schwenk, *Water, The Element of Life*, 1989


*Life is animated water.*
—Vladimir Vernadsky, *The Biosphere*, 1924


*If you want to understand the teaching of water, just drink.*
—Zen saying *in* Alev Lytle Croutier, *Taking the Waters: Spirit, Art, Sensuality*, 1992

# Source Springs: Imbibing Thoreau's Water of Life

### *Henry Thoreau: Immersed Iconoclast*

Of all landscapes, it was the aquatic ones that most inspired Thoreau. "The water on a lake, from however distant a point seen, is always the centre of the landscape," he emphasized. He felt his life to be bounded by water, from his first childhood memories of being taken to visit Walden Pond to his deathbed reflections upon the rain splattering against the windowpanes. Thoreau really did need water for both physical and emotional sustenance: "I should wither up and dry up if it were not for lakes and rivers. I am conscious that my body derives its genesis from their waters." When observing fish swimming, he actually felt himself becoming "amphibious," longing to join them in their watery world.

Thoreau thrilled to the adventure of canoe trips along Maine rivers, reveled in the playfulness of water near his Concord home, and achieved moments of highest spiritual epiphany when drifting in his boat.

But above all, Thoreau was at his very happiest when he was at his very wettest, taking every opportunity to walk along river channels in order to allow his legs to "drink," his body needing a close personal and passionate communion with water: "I must let it soak into me," he once wrote.

### Ecopsychology: Swallowing the Snake and Suckling the Monster to Hatch Water Thoughts

Thoreau, of course, is immensely quotable. Indeed, *Bartlett's Famous Quotations* lists over fifty of his pithy musings and maxims. Everyone has their own particular favorites. For some, it may be one of his philosophical reflections about the human condition: "The mass of men lead lives of quiet desperation," etc.; for others, it is perhaps some of his wondrous prose about nature spirituality: "[A lake] is earth's eye, looking into which the beholder measures the depth of his own nature," etc.; or there are those who are partial to his succinct and opinionated environmental rants: "If some are prosecuted for abusing children, others deserve to be prosecuted for maltreating the face of nature committed to their care." In all of Thoreau's writings there is one account that is quite remarkable for really being quite unlike anything else he ever wrote.

In the spirit of the present volume which focuses on visceral interactions with water—more affairs of the body than those of either the mind or the heart—this particular quotation of Thoreau reveals a side of the normally detached nineteenth century philosopher that comes across as being thoroughly modern in tone and temperament. It is important to realize that one of Thoreau's goals was to achieve a complete immersion and union with the wild. Frequently this took the form of a kind of passionate embrace of nature as lover. Occasionally, though, even this was not enough for Thoreau. Merely existing as part of nature was insufficient; he desperately hungered to absorb its very essence into his own being and to resurrect that wildness deep within. The anecdote recounted in *Walden* about his almost uncontrollable desire to eat a wild woodchuck (this despite being a vegetarian) in order to consume its spirit of wildness, is widely regarded. Almost unknown, however, is his bout of stomach sickness brought about by consuming infested waters. Here, in Thoreau's rambling, nearly nonsensical, stream-of-consciousness prose from August 17, 1851, we encounter an integration with nature whose description engrosses while at the same time it repulses.

> Ah, the very brooks seem fuller of reflections than they were! Ah, such provoking sibylline sentences they are! The shallowest is all at once unfathomable. How can that depth be fathomed where a man may see himself reflected? The rill I stopped to drink at I drink in more than I expected. I satisfy and still provoke the thirst of thirsts. Nut Meadow Brook where it crosses the road beyond Jenny Dugan's that was. I do not drink in vain. I mark that brook as if I had swallowed a water snake that would live in my stomach. I have swallowed something worth the while. The day is not what it was before I stooped to drink. Ah, I shall hear from that draught! It is not in vain that I have drunk. I have drunk an arrowhead. It flows from where all fountains rise.

> How many ova have I swallowed? Who knows what will be hatched within me? There were some seeds of thought, methinks, floating in that water, which are expanding in me. The man must not drink of the running streams, the living waters, who is not prepared to have all nature reborn in him,—to suckle monsters. The snake in my stomach lifts his head to my mouth at the sound of running water. When was it that I swallowed a snake? I have got rid of the snake in my stomach. I drank of stagnant waters once. That accounts for it. I caught him by the throat and drew him out, and had a well day after all. Is there not such a thing as getting rid of the snake, which you have swallowed when young, when thoughtless you stooped and drank at stagnant waters, which has worried you in your waking hours and in your sleep ever since, and appropriated the life that was yours? Will he not ascend into your mouth at the sound of running water? Then catch him boldly by the head and draw him out, though you may think his tail be curled about your vitals.

What is going on here? Of course, the medical explanation is no doubt simply a bout of giardia or cryptosporidia. But can this explain Thoreau's overly emphatic prose? Is there a life lesson for us here, other than the mundane "be careful where or what you drink"?

Yes, indeed, there is an instructive message that can be extracted from Thoreau's hallucinatory prose. He appears to be offering forewarning about the price that must be paid before we can enable nature to be reborn within us. The assimilation and then gestation of true wildness, allowing the breeding of internal "monsters," may be painful, but it is also a process that he acknowledges is ultimately "something worth the while." For it is these same monsters that are rebirthed or transformed into the thoughts that foster our sense of oneness with the Earth from which we have become dissevered. As Joan Halifax writes in *The Fruitful Darkness: Reconnecting with the Body of the Earth*, "Wilderness lives within us as well—in the beating of our blood, in the wild root of our imagination." Birthing, and possibly more importantly, nurturing or "suckling," such monsters of imagination, requires sacrifice. One must venture into the darkness, in Thoreau's sense here, absorbing the stomach-churning demons of wildness, to seek initiation into a more complete relationship with nature. This is the path to true "wisdom," a word, interestingly, originating from the Latin and Hebrew verbs "to taste." Thoreau seems to be telling us that it is not merely enough to passively drink; rather one must abandon caution and gulp deeply in order to actively suck the outside within.

Thoreau's identification of his wildness monster as being a snake is intriguing and illuminating. In Native American shamanism, the snake is the symbol of transmutation, rebirth and ascension, perhaps most powerfully expressed as the famous Serpent Mound in Ohio, where a snake holds the earth-as-an-egg in its mouth. In such a tradition, on a personal energy level, passion and procreation are attributes; on an emotional level, healing and dreams are important; and on a spiritual level, wisdom, wholeness and connection to the Great Spirit are significant. As Jamie Sams and David Carson write: "Snake . . . come crawling/ There's fire in your eyes/ Bite me, excite me/ I'll learn to realize/ The poison transmuted/ Brings eternal flame/ Open me to heaven/ To heal me again."

The healing–snake imagery also shows up as the Greek god Hermes, father of alchemy, in the form of the caduceus used to represent modern medicine. In the ancient Orphic mysteries, Dionysus, the world creator, was himself born out of the cosmic egg around which a serpent spirals. In Judeo-Christian myths, the serpent symbolizes awareness and change in Eden. And finally, and perhaps most significantly, the snake circling back to consume its own tail, the Worm Ouroboros in early paganism, represents the never-ending cyclical nature of death and rebirth on an epic scale.

The imagery of the conjoined mouth and snake figures prominently in this particular quotation from Thoreau. But despite the benefits of having nature reborn within, Thoreau also stresses the need to release that nature again after it has performed its work within you; in his words, by freeing the coiled grasp of the snake from around your "vitals" and then drawing it out through your mouth. There is a need, he seems to be implying, for personal healing, for a recovery from the sometimes too powerful embrace that nature can enact. By being able to step back from the grip of nature, it may then be possible to better see our place within it, benefiting from the distanced perspective, yet never losing sight or memory of the proximity of that fateful consumption of wildness that inspired us. This may be what theologian Matthew Fox means when, referring to creation spirituality, he says that "if we do not honor the animal within, we will not understand how human we are."

The to-and-fro, drug-like experience of imbibing and then recuperating from wildness calls to mind to a quote from Carlos Castaneda that Halifax uses in her book: "Only if one loves this earth with unending passion can one release one's sadness." This concept perfectly encapsulates the modern message behind the developing field of ecopsychology; i.e. one must heal oneself while at the same time striving to heal the earth.

Finally, it is critical to pay attention to Thoreau's suggestion as to the means for bringing about this process of self- and world-restorative healing. He instructs us in this passage to become associated with the "sound of running water," for, as he implores elsewhere, we should all "share the happiness of the river," for s/he "who hears the rippling of the rivers will not utterly despair of anything."

*Time is but a stream I go a-fishing in.*
—Henry Thoreau, *Walden*

# Epilogue

## *STIRRED NOT SHAKEN*

**January, 1854:** *It was as if he had no real choice in the matter; as if somehow the pull had been there ever since he had left the watery womb. After departing from the house in town that morning, he had spent a pleasant and profitable day—walking, observing, recording. Frequently, however, he had found it impossible to completely ignore the mysterious pull, though he had done his best to bury it deep in his mind while focused on detailing the complicated patterns of water currents and ice formations in the river. But his legs, it seemed, had had a mind of their own. For, with a surprised start, he realized that all of a sudden, after wandering about lost in thought, there he was at the site of his old cabin beside the lake. By happenstance or by purpose? he wondered. Smiling ruefully, he finally succumbed, allowing the lake to pull him to herself like a giant magnet, or perhaps a lonely friend.*

*It was as if he had returned to his one true home. It was always such. No matter how far he wandered, it was this particular pond—the "blue navel" of the world, he had once written—around which he circulated. Her umbilical reach extended far. Being "the distiller of celestial dews," the pond's evaporating water seemed to drift out to wherever he was, beckoning, luring—no, that was too passive. What was it? Yes—sucking; that's right, sucking him back. It was as if he, a wayward water droplet, needed to be periodically sucked back to the bosom of the lake source for sustenance.*

*Crouching on the ice a short distance from the shore, with one hit of his fist he broke through the thin covering formed over the fishing hole. Rolling up his sleeves, he plunged his hands and arms deep into the fissure, opening and closing his fingers as if trying to grab a handful of the liquid. Until, with numbness approaching, he formed a cup and drew his hands up high and, throwing back his head and opening his mouth, let the frigid water pour down over his face, bathing him both inside and out.*

117

*Refreshed in body, renewed in spirit, he took out his pencil and notebook and, sitting down on his satchel, returned to his job of recording, for he had just noticed a new phenomena that intrigued him:*

The water on Walden has been flowing into the holes cut for pickerel and others. It has carried with it, apparently from the surface, a sort of dust that collects on the surface, which produces a dirty or grayish-brown foam. It lies sometimes several feet wide, quite motionless on the surface of the shallow water above the ice, and is very agreeably and richly figured, like the hide of some strange beast—how cheap these colors in nature!—parts of it very much like the fur of rabbits, the tips of their tails. I stooped to pick it up once or twice,—now like bowels overlying one another, now like tripe, now like flames, i.e., in form, with the free, bold touch of Nature. One would not believe that the impurities which thus color the foam could be arranged in such pleasing forms. Give any material, and Nature begins to work it up into pleasing forms.

*Closing his notebook, he reflected how it was a constant amazement that no matter how many times he visited the shores, no matter how much he carefully explored the nature there, this particular pond always offered ever more novel details worthy of his continued and close study. Finished, satisfied, he stood up. It was then that he noticed the two individuals on the ice in the distance, at the other end of the pond, one standing still, the other lying down . . .*

**January, 1998:** *The two men had had a wonderful morning walking around pond. For one, a nearby resident, it was a frequent return to a site of profound beauty and historical importance that he never tired of, either when alone or especially when sharing it with guests. For the other, a visitor, it was a new pilgrimage to a source spring of thought that had motivated and guided many of his actions and dreams.*

*They had quickly walked past the replica cabin at the edge of the parking lot, seeking communion with the reality of the pond itself. There, they had lingered at the site of the original cabin, trying to ignore the incongruous interpretive sign instructing about the man who had once lived there. More interesting was the nearby pile of stones added to over the years by other pilgrims as a sort of dynamic, growing monument honoring the scribe. Luckily the two men had had the site to themselves and were thus able to absorb a sense of spirit of the place free from the hordes of tourists and recreationists that clog the shorelines during the summer months.*

*Continuing their walk along the shore path, past where restoration work has had to be implemented in order to prevent further severe erosion, they found the gap in the fence that allowed them to escape onto the surface of the ice. The feeling of freedom and* **joy** *was immediate. The two men slid and spun about in play on the dark ice between the strange patches of brown foam that had collected here and there on the surface. Soon, finding themselves farther away from the shore than they had initially planned, exhilaration was replaced by momentary fear as the newly formed ice settled under their weight with a resounding thump accompanied by a spiderweb network of radiating cracks. Not without trepidation, the two men gingerly made their way back toward the thicker ice located along the edge of the shore, shaken a little by their* **adventure**.

*After several minutes of walking along together in silence and* **contemplation**, *the visitor suddenly dropped to the surface of the ice and, lying upon his back, spread his arms and legs out as if making a snow angel. He was satisfying an immediate urge he explained to his shocked companion. He simply needed to be feel the surface of the ice, to make* **contact** *with the memory of the sage who had once walked upon it. "Heaven is under our feet as well as over our heads," he quickly rattled off the famous quotation. The visitor needed to touch that icy heaven, to press his hands hard against its frigid surface, to feel his body heat melt away a small bit of the divine ice, to lick its essence from his hands, to make deep contact with the spirit of the place and of the man.*

*While the visitor engaged in his immersion, the other man, the resident, looked over the prone body of his friend and into the distance. There, at the far end of the pond, near where they had walked earlier, he could see a small, solitary figure dressed in a long coat, standing motionless on the surface of the ice. The moment, though brief, as his friend would later confirm, seemed for the resident to nevertheless expand beyond its normal temporal bounds. It was as if the stream of time had suddenly reached an eddy of some sort. The distance was too great to make out any details, only that after a few more moments, the figure turned and walked to the shore, disappearing into the trees.*

*Several hours later, the two men, continuing their tour of the environs, made their way to the town cemetery. The snow was covered with a thick crust of ice—a result of a warming spell and rain several days before. With baby steps, they slowly traversed the treacherous graveyard surface to the "Author's Ridge." Tightly gripping the handrail, each wishing for a much needed pair of crampons, the two ascended to the ridge top as if intrepid mountaineers approaching a summit. There, past Nathaniel, past Ralph Waldo, and past Louisa May, they reached Henry.*

*After the customary few minutes of silent acknowledgment and mindful contemplation, the two men, simultaneously infused by uncontrollable boyish prankishness,*

119

*began to wrestle one another atop the icy edge, over the hallowed ground. Slipping and sliding, laughing and lunging, each in mock battle tried to push the other off. Then, with a simultaneous slip, they both skidded down the icy slope, bouncing off the tombstones of the less famous, like steel balls in a giant pinball machine. The resulting jumbled mass of interlocked arms and legs, backpacks and camera bags belonging to the two adult men would have drawn quizzical stares had there been anyone else there to stare. As it was, only the back of Henry's small tombstone, silhouetted large against the winter sun (his "morning star") as he had himself been in life, looked down upon the two men . . . in understanding . . . in gratitude . . . and in blessing.*

*Now comes good sailing . . .*
—among the last words spoken by master
hydrophile, Henry David Thoreau, on his
deathbed, after watching the furrows made by
rain tracing their way down the windowpane


*Water . . .*
—the very last word spoken by poet and ardent
aquatic sensualist, Rupert Brooke, while dying
of blood-poisoning in the Aegean on his
way to Gallipoli

# References

Prologue: Aponiptein, Dikaion Touton

*The Holy Bible. King James Version.* World Book Publ. 1995.

Wroe, Ann. *Pilate: The Biography of an Invented Man.* Vintage. 2000.

Introduction: Ripple Effects

Anon. *The Literary Cat.* Running Press. 1990.

Ball, Philip. *H₂O: A Biography of Water.* Weidenfield & Nicholson Publ. 1999.

Beardsley, John. "Kiss nature goodbye." *In What is Nature Now?* Harvard Design Magazine 10: 60-67. 2000.

Carson, Rachel. *The Sense of Wonder.* Harper and Row. 1956

Finnegan, William. "Letter from Bolivia: Leasing the rain; the world is running out of water, and the fight to control it has begun." *The New Yorker Magazine,* April 2002

Fischesser, Bernard, Marie-France Dupuis-Tate and Hans Silvester. *The Beauty and Mystery of Water.* Harry N. Abrams Inc. 2001.

France, Robert L. (Ed.) *Reflecting Heaven: Thoreau on Water.* Houghton Mifflin Publ. 2001.

France, Robert L. (Ed.) *Profitably Soaked: Thoreau's Engagement with Water.* Green Frigate Books. 2003.

123

France, Robert L. (Ed.) *Handbook of Water Sensitive Planning and Design.* Lewis Publ. 2002.

France, Robert L. (Ed.) *Facilitating Watershed Management: Practical Approaches and Case Studies.* Rowman & Littlefield Publ. 2004.

Illich, Ivan. *H₂O and the Waters of Forgetfulness.* Marion Boyars. 1986.

Lord Selborne. *The Ethics of Freshwater Use: A Survey.* World Commission on the Ethics of Scientific Knowledge and Technology. UNESCO. 2000.

Rodgers, Katherine M. *The Cat and the Human Imagination: Feline Images From Bast to Garfield.* University of Michigan Press. 2000.

Van Alen Institute. 2001. *Architecture + Water.* Van Alen Institute Report 9: May 2001.

Van Engen, Hans, Dietrich Kampe and Sybrand Tjallingii. *Hydropolis: The Role of Water in Urban Planning.* Backhuys Publ. 1995.

Wetzel, Robert. "Salvaging our water commons: greatest threat to humanity." *SIL Annual Circular.* January 2003.

Part 1. Plunging in: Introduction

Chapter One
Main Streams: The Cultural History of Water

Alexander, Olaf. *Living Water: Viktor Schaubergen and the Secrets of Natural Energy.* Gateway. 1990.

Askeland, Jon and Alvaro Ramirez. "Water images in Latin American cinema." *In The Role of Water in History and Development.* Univ. Bergen Conf., August 2001.

Bachelard, Gaston. *Water and Dreams: An Essay on the Imagination of Matter.* Dallas Inst. Humanities and Culture. 1999.

Brettell, Richard R. *French Impressionism.* Abrams Publ. 1987.

Butzer, Karl W. Early *Hydraulic Civilization in Egypt: A Study in Cultural Ecology.* University of Chicago Press. 1976.

Campbell, Craig. *Water in Landscape Architecture*. Van Nostrand Reinhold Publ. 1978.

Christian, Roy. *Well-dressing in Derbyshire*. Derbyshire Country Ltd. 1991.

Croutier, Alev Lytle. *Taking the Waters: Spirit, Art, Sensuality*. Abbeville Press. 1992.

Denver, Bernard. *The Impressionists at First Hand*. Thames and Hudson. 1987.

Douglas, Kate. "Taking the Plunge. Eve's Watery Origins." *New Scientist Magazine* 2266: 29-33, 2000.

Emoto, Masuru. *Messages From Water*. HADO. 2000.

Evans, Mark. "Ancient skull origin in doubt." *The Boston Globe*, August 2002.

Fischesser, Bernard, Marie-France Dupuis-Tate and Hans Silvester. *The Beauty and Mystery of Water*. Harry N. Abrams Inc. 2001.

France, Robert L. *Limnophilia: The Lure of Lakes. Pilgrimage and the Design of Landscapes and Experience*. In prep.

Fratino, Umberto, Antonio Petrillo, Attilio Petruccioli and Michele Stella. *Landscapes of Water: History, Innovation and Sustainable Design, Vols. I & II*. Uniongrafica Corcelli Editrice. 2002.

Fujiwara, Chris. "The good earth: Joris Ivan's elemental cinema." *The Boston Phoenix*, April 2002.

Gogliotta, Guy. "7-million-year-old skull extends human record." *The International Herald*, July 2002.

Guillerme, Andre E. *The Age of Water: The Urban Environment in the North of France, A.D. 300-1800*. Texas A & M University Press. 1983.

*H₂O–The Mystery, Art, and Science of Water*. Web-site witcombe.sbc.edu/water/introduction.html.

Haland, Evy Johanne. "'Let it rain', or 'rain, conceive': rituals of magical rain-making in ancient Greece, a comparative approach." *In The Role of Water in History and Development*. Univ. Bergen Conf., August 2001.

Hartley, Dorothy. *Water in England*. Macdonald and James Publ. 1978.

Higginbotham, James. Piscinae: *Artificial Fishponds in Roman Italy*. Univ. North Carolina Press. 1997.

Hill, Charles H. *The Group of Seven: Art for a Nation*. National Gallery of Canada. 1995.

Ingpen, Robert and Philip Wilkinson. *Encyclopedia of Mysterious Places: The Life and Legends of Ancient Sites Around the World*. Viking Studio Books. 1990.

International Water History Association. Web-site iwha.net.

Joshi, Deepa and Ben Fawcett. "Water, Hindu mythology and an unequal social order in India." *In The Role of Water in History and Development*. Univ. Bergen Conf., August 2001.

King, Angela and Susan Clifford. (Eds.) *The River's Voice: An Anthology of Poetry*. Chelsea Green. 2000.

Kunzig, Robert. "La Marmotta: Lost civilization and underwater stone age city." *Discover Magazine*, November 2002.

Leakey, Richard E.F. and Robert Lewin. *People of the Lake*. Doubleday. 1978.

Litton, R.B. and Robert J. Tetlow. *Water and Landscape: An Aesthetic Overview of the Role of Water in the Landscape*. Water Information Center. 1974.

Magnusson, Roberta J. "Water and wastes in medieval towns." *In The Role of Water in History and Development*. Univ. Bergen Conf., August 2001.

Marks, William E. *The Holy Order of Water: Healing Earth's Waters and Ourselves*. Bell Pond Books. 2001.

Morgan, Elaine. *The Aquatic Ape*. Stein and Day. 1982.

Morgan, Elaine. *The Aquatic Ape Hypothesis*. Souvenir Press. 2001.

Namfe, C.M. "The Lozi water tradition." *In The Role of Water in History and Development*. Univ. Bergen Conf., August 2001.

Newlands, Ann. The Group of Seven and Tom Thompson. *An Introduction*. Firefly Books. 1995.

Pechere, Rene. *With the Running Stream*. Eternit. 1963.

Ryan, William and Walter Pitman. *Noah's Flood: The New Scientific Discoveries About the Event that Changed History*. Touchstone Press. 1998.

Schwenk, Theodor. *Sensitive Chaos: The Creation of Flowing Forms in Water and Air*. Rudolf Steiner Press. 1999.

Schauberger, Viktor. *The Water Wizard*. Gateway. 1995.

Stecker, Hardy. "The runnel: form and function in Persian, Islamic, and Mughal garden design." *GSD paper*, 2002.

Steinberg, Theodor. *Nature Incorporated: Industrialization and the Waters of New England*. University of Massachusetts Press. 1991.

Talmadge, Eric. "Hell Valley's primates: monkeys go ape in hot spa heaven." *The Japan Times*, June 2002.

Tvedt, Terje. *A Journey in the History of Water: Part III. The Myths*. Documentary film. Univ. of Bergen. 2001.

Van Leeuwen, Thomas. *The Springboard in the Pond: An Intimate History of the Swimming Pool*. MIT Press. 1999.

Vicky, Khasandi Inviolata. "'Of frogs' eyes and cows drinking water': the portrayal of water in the folklore of the Kabras of western Kenya." *In The Role of Water in History and Development*. Univ. Bergen Conf., August 2001.

Wendt, Herbert. *The Romance of Water*. Hill and Wong Inc. 1971.

Wescoat, James L., Jr. "Beneath which rivers flow: Water, Geographic imagination and sustainable landscape design." In Fratino, Umberto, Antonio Petrillo, Attilio Petruccioli and Michele Stella. *Landscapes of Water: History, Innovation and Sustainable Design*. Vols. I & II. Uniongrafica Corcelli Editrice, 2002.

Wescoat, James L., Jr. and Joachim Wolschke-Bulmahn. *Mughal Gardens: Sources, Places, Representations, and Prospects*. Dumbarton Oaks. 1996.

Wylson, Anthony. *Aquatecture: Architecture and Water.* The Architectural Press. 1986.

Yun, Zheng Xiao. "Water culture as ethnic tradition and sustainable development of the Tai peoples of China." *In The Role of Water in History and Development.* Univ. Bergen Conf., August 2001.

## Chapter Two
### Undercurrents: Sources of Aquatic Anxiety

Arnett, Allison. "Water, water everywhere." *The Boston Globe,* May 15, 2002.

Blatter, Joachim and Helen Ingram. *Reflections on Water. New Approaches to Transboundary Conflicts and Cooperation.* MIT Press. 2001.

Clarke, Robin. *Water: The International Crisis.* MIT Press. 1993.

Cowan, James. *A Mapmaker's Dream.* Shambhala Publ. 1996.

De Villers, Marq. *Water.* Stoddart Publishing. 1999.

Dillard, Annie. *For The Time Being.* Vintage. 1999.

Finnegan, William. "Letter from Bolivia: Leasing the rain; the world is running out of water, and the fight to control it has begun." *The New Yorker Magazine,* April 2002

Fischesser, Bernard, Marie-France Dupuis-Tate and Hans Silvester. *The Beauty and Mystery of Water.* Harry N. Abrams Inc. 2001.

France, Robert L. *Wetland Design: Principles and Practices for Landscape Architects and Land-Use Planners.* W.W. Norton. 2003.

France, Robert L. (Ed.) *Handbook of Water Sensitive Planning and Design.* Lewis Publ. 2002.

Gleick, Peter H. *Water in Crisis: A Guide to the World's Freshwater Resources.* Oxford University Press. 1993.

Naiman, Robert J., John J. Magnuson, Diane M. McKnight and Jack A. Standford (Eds.). *The Freshwater Imperative: A Research Agenda.* Island Press. 1995

National Geographic Staff. "Water: The Power, Promise, and Turmoil of North America's Fresh Water." *National Geographic Magazine* Special Issue. 1993.

Overwater, Alice. *Water: A Natural History*. Perseus Books. 1996.

Postel, Sandra. *Last Oasis: Facing Water Scarcity*. W.W. Norton. 1992.

Robinson, Sandra, Dennis Nelson, Susan Higgins and Michael Brody. *Water, A Gift of Nature: The Story Behind the Scenery*. KC Publications. 1993.

Schwenk, Theodor. *Water, The Element of Life*. Anthroposophic Press. 1989.

Swanson, Peter. *Water, the Drop of Life. Companion to the Public Television Series*. NorthWood Press. 2001.

Woodward, Colin. *Ocean's End. Travels Through Endangered Seas*. Basic Books. 2000.

Wroe, Ann. *Pilate: The Biography of an Invented Man*. Vintage. 2000.

Chapter Three
Head Waters, Water Balms: Liquid Solace

Abrams, David. *The Spell of the Sensuous*. Vintage Books, 1996.

Anderson, Lorraine. "Wilderness in the blood." In Adams, C. (Ed.) *The Soul Unearthed. Celebrating Wildness and Personal Renewal Through Nature*. Tarcher/Putnam. 1996.

Becker, Bruce and Andrew Cole. *Comprehensive Aquatic Therapy*. Healing Press. 1997.

Calahan, William. "Ecological groundeness in gestalt therapy." In Rozak, T., M. E. Gomes and A. D. Kanner (Eds.) *Ecopsychology. Restoring the Earth, Healing the Mind*. Sierra Club Books. 1995.

Clinebell, H. *Ecotherapy: Healing Ourselves, Healing the Earth*. Fortress Press. 1996.

Croutier, Alev Lytle. *Taking the Waters: Spirit, Art, Sensuality*. Abbeville Press. 1992.

Cumes, D. Inner Passages, *Outer Journeys: Wilderness, Healing, and the Discovery of Self.* Llewellyn, 1998.

Deming, H.G. *Water, the Fountain of Opportunity.* Oxford University Press. 1975.

Faber, Thomas. *On Water.* The Ecco Press. 1994.

Feldhaus, Anne. *Water and Womanhood.* Oxford Univ. Press. 1995.

France, Robert. *Water-Logged-In, An Ebook: A Dynamic Library of Aquatic Quotations from Thoreau's Descendants.* Web-site: water-logged-in.com.

Gill, Alexandra. "Paddling out to purity." *The Globe and Mail*, September 2000.

Gomes, Mary E. and Allen D. Kanner. "The rape of the well-maidens: feminist psychology and the environmental crisis." *In* Rozak, T., M. E. Gomes and A. D. Kanner (Eds.) *Ecopsychology. Restoring the Earth, Healing the Mind.* Sierra Club Books. 1995.

Haland, Evy Johanne. "'Let it rain', or 'rain, conceive': rituals of magical rain-making in ancient Greece, a comparative approach." *In The Role of Water in History and Development.* Univ. Bergen Conf., August 2001.

Halifax, Joan. *The Fruitful Darkness: Reconnecting with the Body of the Earth.* Harper Collins, 1993.

Henderson, Karla. "Feminist perspectives, female ways of being, and nature." *In* Driver, B.L. et al. (Eds.) *Nature and the Human Spirit. Toward an Expanded Land Management Ethic.* Venture Publ. 1996.

Janaki, D. "Aadiperukku: water ritual through religious practice." *In The Role of Water in History and Development.* Univ. Bergen Conf., August 2001.

Kidner, David W. *Nature and Psyche: Radical Environmentalism and the Politics of Subjectivity.* State University of New York. 2001.

Macy, Joanna and M. Y. Brown. *Coming Back to Life: Practices to Reconnect Our Lives, Our World.* New Society. 1998.

Marks, William E. *The Holy Order of Water: Healing Earth's Waters and Ourselves*. Bell Pond Books. 2001.

Marrin, West. *Universal Water: The Ancient Wisdom and Scientific Theory of Water*. Inner Ocean Publ. 2002.

Muryn, Mary. *Water Magic: Healing Bath Recipes for both Body and Soul*. Fireside Books. Simon and Schuster. 1995.

Ostray, Judy. "The water cure: take inspiration from the Japanese ritual of the bath." *Natural Home Magazine*, September 2002.

Rozak, Theodore. *The Voice of the Earth: An Exploration of Ecopsychology*. Simon & Schuster. 1992.

Ryrie, Charlie. *The Healing Energies of Water*. Journey Editions. 1999.

Sewall, Laura. *Sight and Sensibility: The Ecopsychology of Perception*. Tarcher/Putnam. 1999.

Sprawson, Charles. *Haunts of the Black Masseur: The Swimmer as Hero*. Pantheon, 1992.

Tvedt, Terje. *A Journey in the History of Water. Part III. The Myths*. Documentary film. Univ. of Bergen. 2001.

Troeller, Linda. *Healing Waters*. Aperture Press. 1998.

Weiss, Harry. *The Great American Water-cure Craze, a History of Hydrotherapy in the United States*. Doubleday. 1967.

Wheelwright, J. H. and L. Wheelwright Schmidt. *The Long Shore: A Psychological Experience of the Wilderness*. Sierra Club Books. 1991.

Wright, Rebekah. *Hydrotherapy in Psychiatric Hospitals*. Harper Collins. 1940.

Woman and Water Conference. *The Bounds Between Women and Water*. Web-site d.umn.eedu/women.water.

Part II. Wetted Appetites: Approaches

Chapters Four–Seven
Adventure, Joy, Contact, Contemplation

Abbey, Edward. *Down the River.* E.P. Dutton Inc. 1982.

Brokaw, Tom, Robert Collins and Roderick Nash, Rodger Drayna, J. Calvin Giddings, John Malo *In* Huser, Verne (Ed.) *River Reflections: An Anthology.* The Globe Pequot Press. 1988.

Buell, Lawrence. *The Environmental Imagination: Thoreau, Nature Writing, and the Formation of American Culture.* Harvard University Press. 1995.

Childs, Craig. *The Secret Knowledge of Water: Discovering the Essence of the American Desert.* Sasquatch Books. 2000.

Deakin, Rodger. *Waterlog: A Swimmer's Journey Through Britain.* Random House. 2000.

Delp, Michael. *Under the Influence of Water: Poems, Essays, and Stories.* Great Lakes Books. 1992.

Delp, Michael. *The Coast of Nowhere: Meditations on River, Lakes and Streams.* Great Lakes Books. 1997.

Duff, Chris. *On Celtic Tides: One Man's Journey Around Ireland by Sea Kayak.* St. Martin's Press. 2000.

Eiseley, Loren. *The Immense Journey.* Random House. 1957.

Eiseley, Loren. *The Night Country.* University of Nebraska Press. 1997.

Faber, Thomas. *On Water.* The Ecco Press. 1994.

France, Robert L. (Ed.) *Reflecting Heaven: Thoreau on Water.* Houghton Mifflin Publ. 2001.

France, Robert L. (Ed.) *Profitably Soaked: Thoreau's Engagement with Water.* Green Frigate Books. 2003.

France, Robert. *Water-Logged-In, An Ebook: A Dynamic Library of Aquatic Quotations from Thoreau's Descendants.* Web-site water-logged-in.com.

Green, Bill. *Water Ice & Stone: Science and Memory on the Antarctic Lakes.* Harmony Books. 1995.

Grey Owl. *A Book of Grey Owl: Selections From His Wild-life Stories.* Macmillan Company. 1964.

Harrigan, Stephen. *Water and Light: A Diver's Journey to a Coral Reef.* Houghton Mifflin. 1992.

Harris, Eddy. *In* Murray, John A. (Ed.) *The River Reader.* The Lyons Press. 1998.

Heat-Moon, William Least. *River-Horse: Across America by Boat.* Houghton Mifflin Company. 1999.

Hemmingway, Lorian. *Walk on Water: A Memoir.* Simon & Schuster. 1998.

Hildebrand, John. *Reading the River: A Voyage Down the Yukon.* Houghton Mifflin. 1988.

Jerome, John. *Blue Rooms: Ripples, Rivers, Pools, and Other Waters.* Henry Holt and Company. 1997.

Kanzantzakis, Nikos. *Japan, China.* Simon and Schuster. 1963.

Lambert, Craig. *Mind Over Water: Lessons on Life From the Art of Rowing.* Houghton Mifflin, 1998.

Lembke, Janet. *Skinny Dipping and Other Immersions in Water, Myth, and Being Human.* Lyons & Burford Publishers. 1994.

Linnea, Ann. *Deep Water Passage: A Spiritual Journey at Midlife.* Simon & Schuster Inc. 1993.

Lopez, Barry. *Desert Notes—Reflections in the Eye of a Raven. River Notes—the Dance of Herons.* Avon Books. 1990.

Middleton, Harry. *Rivers of Memory.* Pruett Publising. 1993.

Mohanraj, Mary Anne. (Ed.) *Aqua Erotica: 18 Stories for a Steamy Bath.* Three Rivers Press. 2000.

Moore, Kathleen Dean Moore. *Riverwalking: Reflections on Moving Water.* Harcourt Brace & Company. 1995.

Nelson, Richard. *The Island Within.* Random House. 1989.

Noyce, Wilfred. *The Springs of Adventure.* John Murray Publ. 1958.

Offutt, Chris. *The Same River Twice: A Memoir.* Penguin Books. 1994.

Patterson, Kevin. *The Water in Between: A Journey at Sea*. Random House. 1999.

Patterson, R.M. *Dangerous River: Adventure on the Nahanni*. Boston Mills Press. 1989.

Peterson, Brenda. *Living by Water: Essays on Land and Spirit*. Northwest Publ. 1990.

Quammen, David. *Wild Thoughts from Wild Places*. Scribner. 1998.

Raban, Jonathan. *Passage to Juneau: A Sea and Its Meanings*. Pantheon Books. 1999.

Ricks, Byron. *Homelands: Kayaking the Inside Passage*. Avon Books. 1999.

Servid, Carolyn. *Of Landscape and Longing: Finding a Home at the Water's Edge*. Milkweed Editions. 2000.

Sprawson, Charles. *Haunts of the Black Masseur: The Swimmer as Hero*. Pantheon, 1992.

Starker, Don. *Paddle to the Arctic: The Incredible Story of a Kayak Quest Across the Roof of the World*. McClelland & Steward Inc.1995.

Stegner, Wallace. *The Sound of Mountain Water*. Doubleday. 1969.

Theroux, Paul. *The Happy Isles of Oceania: Paddling the Pacific*. Ballantine Books. 1992.

Turk, Jon. *Cold Oceans: Adventures in Kayak, Rowboat, and Dogsled*. Harper Perennial. 1998.

Wallach, Jeff. *What the Rivers Says*. Blue Heron Publ. 1996.

Zwinger, Ann. *Run, River, Run: A Naturalist's Journey Down One of the Great Rivers of the West*. Harper & Row. 1975.

Zwinger, Ann and Edwin Way Teale. *A Conscious Stillness: Two Naturalists on Thoreau's Rivers*. Harper & Row, New York. 1982.

Part III. Surfacing: Implementation

Chapter Eight
Reversing the Flow: Human–Water Restorations

Baldwin, A.D., Judith De Luce and Carl Pletsch. (Ed.) *Beyond Preservation: Restoring and Inventing Landscapes*. University of Minnesota Press. 1994.

Barton, Nicholas. *The Lost Rivers of London*. Historical Publications. 1992.

Cassuto, David N. Cold *Running River*. The University of Michigan Press. 1994.

Common Ground. Web-site commonground.org.uk.

Conradin, Fritz. "The Zurich stream daylighting program." *In* Kirkwood, N. and R. France (Eds.) *Brown Fields and Gray Waters: Reclaiming, Remediating and Restoring Post-Industrial and Degraded Landscapes*. Lewis/CRC Press. 2004.

Cronin, John and Robert Kennedy, Jr. *The Riverkeepers. Two Activists Fight to Reclaim our Environment as a Basic Right*. Touchstone Publ. 1999.

Dann, Kevin. *Lewis Creek Lost and Found*. Middlebury College Press. 2001.

EPA. *Top 10 watershed lessons learned*. EPA. 1997.

Findley, James. A. (Ed.) *The Rivers of America*. Web-site co.board.fl.us/ii07700.htm.

Forbes, Peter. *The Great Remembering: Further Thoughts on Land, Soul, and Society*. Center for Land and People; Trust for Public Land Publ. 2001

France, Robert. (Ed.) *Healing Natures, Repairing Relationships: The Restoration of Ecological Spaces and Consciousness*. MIT Press. 2004.

France, Robert. "Las Vegas Wash to Clark County Wetlands Park." *In* France, Robert and Niall Kirkwood (Eds.) *Reclaimed! Recovery Processes and Design Practices for Post-Industrial Landscapes*. Spon. 2004.

France, Robert. "Barn Elms Reservoirs to The Wetland Centre." *In* France, Robert and Niall Kirkwood (Eds.) *Reclaimed! Recovery Processes and Design Practices for Post-Industrial Landscapes*. Spon. 2004.

France, Robert Lawrence. *Still Waters, Lost...Still, Waters Lost: An Environmental Narrative of Sin and Purgatory.* In prep.

Gobster, Paul and Bruce Hull. (Eds.) *Restoring Nature: Perspectives from the Social Sciences and Humanities.* Island Press. 2000.

Gunn, John. "Restoration of the Sudbury mining-impacted ecosystem: starting with its people." *In* Kirkwood, N. and R. France (Eds.) *Brown Fields and Gray Waters. Reclaiming, Remediating and Restoring Post-Industrial and Degraded Landscapes.* Lewis/CRC Press. 2004.

House, Freeman. *Totem Salmon: Life Lessons From Another Species.* Beacon Press. 1999.

Hull, Bruce and David Robertson. "Conclusion. Which nature?" *In* Gobster, Paul and Bruce Hull. (Eds.) *Restoring Nature: Perspectives from the Social Sciences and Humanities.* Island Press. 2000.

Jones, Mary Margaret. "The Crissy Fields saltmarsh restoration." *In* Kirkwood, N. and R. France (Eds.) *Brown Fields and Gray Waters: Reclaiming, Remediating and Restoring Post-Industrial and Degraded Landscapes.* Lewis/CRC Press. 2004.

Jordan, William. "Restoration, community, and wilderness." *In* Gobster, Paul and Bruce Hull. (Eds.) *Restoring Nature: Perspectives from the Social Sciences and Humanities.* Island Press. 2000.

Kadlecik, Laura and Mike Wilson. "Wetlands and wellness at the United Indian Health Services Potawot Health Village." *In* Kirkwood, N. and R. France (Eds.) *Brown Fields and Gray Waters: Reclaiming, Remediating and Restoring Post-Industrial and Degraded Landscapes.* Lewis/CRC Press. 2004.

Mason, Gary. "Water reborn. A story about stream 'daylighting' in the San Fransisco Bay area." *In* Kirkwood, N. and R. France (Eds.) *Brown Fields and Gray Waters: Reclaiming, Remediating and Restoring Post-Industrial and Degraded Landscapes.* Lewis/CRC Press. 2004.

Mathur, Anuradha and Delip da Cunha. *Mississippi Floods: Designing a Shifting Landscape.* Yale Univ. Press. 2001.

Mills, Stephanie. *In Service of the Wild: Restoring and Reinhabiting Damaged Land.* Beacon Press. 1995.

Patchett, James and Gerould Wilhelm. "The ecology and culture of water." *Conservation Design Forum Tech. Paper*, 1999.

Rapp, Valerie. *What the River Reveals: Understanding and Restoring Healthy Watersheds.* The Mountaineers Press. 1997

Rothenberg, David and Martha Ulvaeus. (Eds.) *Writing on Water. A Terra Nova Book.* MIT Press. 2001.

Spirn, Anne Whiston. *The Language of Landscape.* Yale University Press. 1998.

Throop, William. (Ed.) *Environmental Restoration: Ethics, Theory, and Practice.* Humanity Books. 2000.

White, Jonathan. (Ed.) *Talking on the Water: Conversations About Nature and Creativity.* Sierra Club Books. 1994.

Wilder, Thornton. *The Bridge of San Luis Rey.* Avon Books. 1927.

Williams, Jack E., Christopher A. Wood and Michael P. Dombeck. (Eds.) *Watershed Restoration: Principles and Practices.* American Fisheries Society. 1997.

Woodward, Joan. *Waterstained Landscapes: Seeing and Shaping Regionally Distinctive Places.* John Hopkins University Press. 2000.

## Chapter Nine
### Turning the Tide: Rehydration Education

Anon. "When you're here, you're thirsty." *Harpers Magazine*, October 2002.

Brown, Brenda. J. "Harvesting what you can't hold tight. Water—from sustainable architecture to graceful amenity, from dramatic actor to scenic prop." *Landscape Architecture Magazine.* 7/2001.

Brukilacchio, Lisa and Jennifer Hill. "Taking it to the streets: Mystic Watershed Awareness efforts merge with public art." *In* France, Robert L. (Ed.) *Facilitating Watershed Management: Practical Approaches and Case Studies.* Rowman & Littlefield Publ. 2004.

Cavicchi, Elizabeth. "Exploring water: art and physics in teaching and learning with water." *In* France, Robert L. (Ed.) *Facilitating*

*Watershed Management: Practical Approaches and Case Studies.* Rowman & Littlefield Publ. 2004.

Collins, Timothy. "Art, nature, culture and restoration ecology" *In* France, R. (Ed.). *Healing Natures, Repairing Relationships: The Restoration of Ecological Spaces and Consciousness.* MIT Press. 2003.

Crandell, Gina. "Atelier Dreiseitl." *LandForum Magazine.* 2001.

Damon, Betsy and Anne H. Mavor. "Combing art and science: The Living Water Garden." *In* France, Robert L. (Ed.) *Facilitating Watershed Management: Practical Approaches and Case Studies.* Rowman & Littlefield Publ. 2004.

Dannenhauer, Mark. "On performance and water sensitive ecological planning and design." *In* France, R. (Ed.) *Healing Nature, Repairing Relationships: The Restoration of Ecological Spaces and Consciousness.* MIT Press. 2004.

Dreiseitl, Herbert, Dieer Grau and Karl H.C. Ludwig. *Waterscapes: Planning, Building and Designing with Water.* Birkhauser Press. 2001.

Eiseley, Loren. *The Immense Journey.* Random House, New York. 1957.

France, Robert. "Smoky mirrors and unreflected vampires: From eco-revelation to eco-relevance in landscape design." *In* What is Nature Now? *Harvard Design Magazine.* 10: 36-40. 2000.

France, Robert. "Waterscapes, reviewed." *Landscape Architecture Magazine.* 2001.

France, Robert and Francois De Kock. *Water Sensitive Ecological Planning and Design.* Web-page gsd.harvard.edu/watersymp.

Harries, Mags and Lajos Heder. "The touch of water: artists working with water and communities." *In* France, Robert L. (Ed.) *Facilitating Watershed Management: Practical Approaches and Case Studies.* Rowman & Littlefield Publ. 2004.

Illich, Ivan. *$H_2O$ and the Waters of Forgetfulness.* Marion Boyars. 1986.

International Rivers Network. *River of words.* Web-site: irn.org.

International Water History Association. 2001. *The role of water in history and development.* Web-site: iwha.org.

Irland, Basia. *Water Library, Chapter One: A Sculptor's Research into the Phenomenon of Water.* The Univ. New Mexico Art Museum. 2001.

Keepers of the Waters. Web-site keepersofthewaters.org.

North, Peter. "Elevating wetland consciousness: environmental education and art." *In* France, Robert L. (Ed.) *Facilitating Watershed Management: Practical Approaches and Case Studies.* Rowman & Littlefield Publ. 2004.

Poole, Kathy. *Boston's Back Bay Fens: Model of Cultural and Ecological Dynamics.* Brochure and web-site: iath.virginia.edu/backbay/

Rinne, Katherine. *Aquae Urbis Romae: The Water of the City of Rome.* Web-site: iah.viginia.edu/waters/

Roybal, Valerie. "A gathering of waters. Approaching water issues through artful life." *Quantum Magazine.* Univ. New Mexico. 1997.

Stilgoe, John. *Outside Lies Magic: Regaining History and Awareness in Everyday Life.* Walker Press. 1998.

Surfrider Foundation. *From Sea to Summit.* Surfrider.org. 2002.

WaterFire. Web-site: waterfire.com.

## Chapter Ten
### Source Springs: Imbibing Thoreau's Water of Life

France, Robert L. (Ed.) *Reflecting Heaven: Thoreau on Water.* Houghton Mifflin Publ. 2001.

France, Robert L. (Ed.) *Profitably Soaked: Thoreau's Engagement with Water.* Green Frigate Books. 2003.

Halifax, Joan. *The Fruitful Darkness: Reconnecting with the Body of the Earth.* Harper Collins. 1993.

Sams, Jamie and David Carson. *Medicine cards.* St. Martin's Press. 1999.

# Epilogue: Stirred Not Shaken

France, Robert L. *Watermarks: Imprints Upon a Life*. In prep.

# Quotation Index

Carson, Rachel: xxxix

Castaneda, Carlos: 115

Cavicchi, Elizabeth: 98

Cherry-Gerard, A.: 48

Childs, Chris: 55, 61, 63, 66, 67, 88

Clinebell, Howard: 39

Claudel, Paul: 37, 42

Collingwood, R.G.: xiv

Cousteau, Jacques: xxxii, 64, 68

Croutier, Alev Lytle: 110

Cumes, David

# D

Damon, Betsy and Anne Mavor: 101

Dannenhauser, Mark: 104

Da Vinci, Leonardo: x

Deakin, Roger: 61, 66, 67, 68, 70

De Chardin, Pierre Telilhard: 3, 81

De Lamartine, Alphonse: 45

Delp, Michael: 67, 69, 70, 71, 77, 79, 80, 82, 83

Deming, H.G.: 42

Dillard, Annie: 31

Dreiseitl, Herbert: 102

Dreyna, Chris: 50

Duff, Chris: 50, 53, 55, 59, 62, 77, 78, 83

Duncan, David James: 34

# E

# F

# G

# H

# M

# N

# O

# P

# Q

# R

# S

# T

Theroux, Paul: 52, 84

Troeller, Linda: 44

Turk, Jon: 50, 53, 58, 61, 62

# V

Valery, Paul: 69

Van Vechten, Carl: xxxii

Vernadsky, Vladimir: 110

Vitruvius: 2

# W

Wallach, Jeff: 58, 77, 80, 81

WaterFire Providence: 108

Wescoat, James, Jr.: ix

Wetzel, Robert: xxxiv

Wilder, Thorton: 92

Winfrey, Oprah: 2

Williams, Tennessee: 82

Williams, Terry Tempest: 68, 70

Wittgenstein, Ludwig: 9

Wroe, Ann: 16

# Z

Zwinger, Ann: 53, 58, 59, 63, 67, 69, 83, 84

# Acknowledgements

Thanks to Mina France for aiding in manuscript preparation. And thanks to all those who have allowed me the opportunity to experience the liquid world in their presence, and thus be doubly enriched. Particularly, I would like to thank Carolyn Adams, Mark Chandler, Verne Huser, Lewis MacAdams, John Middendorf, Lawrence Millman, Sandra Postel, James Wescoat, and especially Laura Sewall and Herbert Dreiseitl—impassioned lovers of, and enthusiastic dabblers in, waters all—for their thoughtful and inspiring words. This work was funded in part by operating grants through the Harvard Design School.

For more on water in literature, browse *Water-logged-in.com: A Dynamic Library of Aquatic Quotations from Thoreau's Descendants,* an eBook published at:

water-logged-in.com

"These quotations are a potent reminder to get outside, to get dirty, to get wet, and to rejoice in your lived experience. They should catalyze action, spur you to taste the world, and remember that you are part of nature. And when you return form your adventure, you will be changed for the better...Read these quotations. Enjoy these stories of those who came before you. You will be enobled."

*—Robert Abbott, Abbott Strategies*

"While Water-logged-in claims to offer non-fiction quotations as literary solace in an age of information saturation, its true passion flows directly from France's faith in the restorative power of aquatic immersion...In aggregate, the quotations in Water-logged-in bear compelling witness to the centrality of water in any understanding of human nature...Water-logged-in prompts no less than immersion itself, be it of the intellectual or the embodied sort. As the quotations at the heart of this collection suggest, however, the best sort of dips will be both."

*—Cheryl Foster, University of Rhode Island*

"Robert France has brought luminous water voices together into the most dynamic library imaginable. Enjoy the adventure, the joy, the contact and contemplation, and the mystery of water."

*—Grant Jones, Jones & Jones Architects and Landscape Architects*

Water-logged-in.com was prepared as a reader for a course at Harvard Universtiy, *Ecopsychology: Human–Nature Relationships.*

149

Thoreau knew that water can change the focus of the spirit. The directionality of moving water is powerful, yet creates a peace. With water we can visit the world, learn to move with it, and be a part of its moments. With time a knot in water slowly loosens and untangles. Similarly, water can untangle one's heart and mind. No one who connects with the world of water can remain unmoved by it. Water has the ability to sublimate our energies into the spiritual realm. Thankfully, Robert France has perfectly captured and consolidated the Master's timeless work to bring Thoreau's essence alive.

> ✍ *John Middendorf,* professional river guide, international big wall climber, and equipment designer

As Robert France shows us in this exhilarating and thought-provoking book, Thoreau's visionary transcendentalism has its boots deep in the mud of experience; and it is this embodied vision that makes him so significant a seer for our dissociated times. Thoreau's deceptively gentle style belies a profound radicalism that invites us to cast off from our islands of imposed intellectual order and to immerse ourselves fully in nature's vitality and variety, inspiring us toward a fuller engagement with the natural world.

> ✍ David Kidner, professor at Nottingham Trent University and author of *Nature and Psyche. Radical Environmentalism and the Politics of Subjectivity*

Again I am impressed with how alive Thoreau was – and after one hundred and forty years – still is. How still and focused he could be. How much of his communion, recreation and travel involved being with ponds, lakes, rivers, and the ocean. How often wet he was! How in love with the world he was. And I thank Professor France's stimulating presentation "Profitably Soaked" for this gift.

> ᔐ    J. Parker Huber, author of *The Wildest County: A Guide to Thoreau's Maine*

In terms of intervention, what knowledge, per chance would Thoreau have us bring to a cycle of culturally conditioned experience and the actions that arise from such conditioned states? And what knowledge would we descendants of Thoreau, the contemporary lovers of land and water, have to bring to the conscious construction of our sensibilities? My guess is that Thoreau would emphatically suggest knowledge born out of bodily experience with the natural world. In the present compilation of quotations, Thoreau charges us to walk, listen, look, and immerse ourselves. This is how we know. It is only with experience, with sincere immersion in the sensible world, that our bodies begin to know—and thus to inform our every step.

> ᔐ    Laura Sewall, professor at Prescott College and author of *Sight and Sensibility. The Ecopsychology of Perception*

One would expect that an inveterate pondside dweller such as Henry Thoreau would have something to say about water. But who would have thought that he could be so ecstatic about the subject. And so eloquent. A fine assembly of Thoreau's thoughts.

> ᔐ    John Hanson Mitchell, editor of *Sanctuary Magazine* and author of *Walking Towards Walden, A Pilgrimage in Search of Place*

# GREEN FRIGATE BOOKS

## "THERE IS NO FRIGATE LIKE A BOOK"

*Words on the page have the power to transport us, and in the process, transform us. Such journeys can be far reaching, traversing the landscapes of the external world and that within, as well as the timescapes of the past, present and future.*

**Green Frigate Books** is a small publishing house offering a vehicle—a ship—for those seeking to conceptually sail and explore the horizons of the natural and built environments, and the relations of humans within them. Our goal is to reach an educated lay readership by producing works that fall in the cracks between those offered by traditional academic and popular presses.